Mathematics and Philosophy at the Turn of the First Millennium

At the turn of the first millennium, scientific and philosophical knowledge was far from dormant. Arithmetic, with its diverse calculation techniques and number theory, served as a bridge to philosophy, theology, and the study of the physical world. Even something as simple as a series of multiplication tables could unlock a profound knowledge of both the divine realm and natural phenomena. Such is the case with Abbo of Fleury's *Commentary on the Calculus*.

Mathematics and Philosophy at the Turn of the First Millennium sheds light on Abbo's original philosophical system anchored in two central doctrines, which serve as a compass to navigate it: the theory of unity (henology) and the theory of composition. Yet, the *Commentary on the Calculus* covers much more. The present study, thus, explores an eclectic range of topics – from water clocks to barleycorns, constellations to human voice, synodic month to the human lifespan, and numbers to God. Abbo's work is an ambitious attempt to tie together the study of both the visible and invisible realms, what can be measured and what cannot, what can be quantified and what exceeds quantification.

Scholars and students of the history of philosophy and mathematics will be introduced to a pivotal figure from an often overlooked era. They will be provided with fresh insights into the spread of Neopythagorean doctrines in the early Middle Ages, as they learn how these ideas were transmitted through arithmetic texts and harmonised with theology and natural philosophy. They will also get to know the medieval fraction system and calculus practices.

Clelia V. Crialesi is a Marie Skłodowska-Curie Fellow at SPHERE-CNRS (France). Formerly, she was an FWO Research Fellow at KU Leuven (Belgium) and a Mellon Fellow at PIMS (Canada). Her research focuses on premodern mathematical thought, with publications ranging from Boethian number theory to Euclidean geometry in the late medieval continuum debate.

GLOBAL PERSPECTIVES ON THE HISTORY OF NATURAL PHILOSOPHY

Fostering a new approach to the study of the history of natural philosophy this series aims to expand the discussion on natural philosophy cross-culturally and comparatively by focusing on the philosophical reasoning about nature developed particularly, but not exclusively, in three main cultural settings: Europe, the Middle East, and China. One of the main focal points of *Global Perspectives on the History of Natural Philosophy* is the interplay between philosophical and scientific concepts, stances, and problems arising from the premodern consideration of nature, broadly considered. Accordingly, the series provides a cutting-edge framework in which natural philosophy can be considered from new philosophically meaningful angles. Acknowledging the historical interweaving of philosophy and science of nature, the series publishes monographs and edited volumes dealing with the history of natural philosophy from three methodological perspectives: philosophical analysis, historical reconstruction, and comparative studies. Submitted manuscripts may either examine authors and issues from a specific philosophical tradition or engage comparatively with patterns and problems shared by different cultural settings.

Theories of Colour from Democritus to Descartes
Edited by Véronique Decaix and Katerina Ierodiakonou

Dante's Visions
Crossing Sights on Natural Philosophy, Theory of Vision, and Medicine in the Divine Comedy and Beyond
Edited by Cecilia Panti and Marco Piccolino

Mathematics and Philosophy at the Turn of the First Millennium
Abbo of Fleury on *Calculus*
Clelia V. Crialesi

Mathematics and Philosophy at the Turn of the First Millennium

Abbo of Fleury on *Calculus*

Clelia V. Crialesi

LONDON AND NEW YORK

First published 2025
by Routledge
4 Park Square, Milton Park, Abingdon, Oxon OX14 4RN

and by Routledge
605 Third Avenue, New York, NY 10158

Routledge is an imprint of the Taylor & Francis Group, an informa business

British Library Cataloguing-in-Publication Data
A catalogue record for this book is available from the British Library

ISBN: 978-1-032-64345-8 (hbk)
ISBN: 978-1-032-64346-5 (pbk)
ISBN: 978-1-032-64347-2 (ebk)

DOI: 10.4324/9781032643472

Typeset in Sabon
by KnowledgeWorks Global Ltd.

Contents

Figures

Acknowledgements

This book represents the culmination of extensive research conducted within the framework of an agreement between the Doctoral School in Philosophy at the Universities of Rome Tor Vergata and Roma Tre, and École Doctorale 472, Religions et Systèmes de Pensée, at the École Pratique des Hautes Études in Paris. The realization of this work was made possible through a three-year grant provided by the affiliated Italian universities, complemented by the support of the 'Bando Vinci', the thesis mobility grant from the Italian-French University. I extend my gratitude to other institutions, particularly the Laboratoire des Études sur les Monothéismes (LEM) and the Institut de Recherche et d'Histoire des Textes (IRHT) of the Centre National de la Recherche Scientifique (CNRS), for facilitating and enriching this research endeavour. I heartfeltly thank Cecilia Panti and Irene Caiazzo, who acted as my doctoral supervisors. Their commitment to academic excellence has been instrumental in the successful completion of my doctoral dissertation, from which the present book stems. I would like to express my deepest gratitude to my partner, Nicolò. His dedication to creating all the diagrams and figures featured in this book has not only added a visual dimension to the research but has also elevated the overall quality of the presentation.

Introduction

A spectre looms over the 10th century, often considered the darkest period of the Middle Ages: the prevailing modern notion suggests that little of philosophical and scientific significance occurred at the turn of the first millennium. Consequently, what transpired during this era is often casually dismissed as an unproductive reworking of a limited selection of late-antique and early medieval sources. Not only is this report inaccurate, but it also fosters the misconception that science and philosophy were relatively dormant between the more renowned 'renaissances' of the Carolingian and the 12th century.[1] Studies focused on Gerbert of Aurillac (ca. 950–1003) have played a crucial role in challenging and undermining this belief.[2] The present book builds upon this research trajectory, ultimately dismantling such misconceptions and reshaping our comprehension of the 10th century. It does so by concentrating on a central figure of that time, Abbo of Fleury (ca. 945–1004). While Abbo's name may evoke some recognition, insufficient attention has been directed towards this prominent figure in early medieval European history of philosophy and science. As a monk, abbot, and master at Fleury and Ramsey, Abbo was well-versed in all the liberal arts, with a particular emphasis on mathematics. His expertise spanned – yet was not limited to – astronomy, computus, music, arithmetic, and calculus. In this context, the term 'calculus' is employed to denote a specific form of practical arithmetic that involves arithmetical operations, ratios, and proportions, encompassing both integers and fractions.[3] This book sheds light on the multifaceted contributions of Abbo in both philosophy and mathematics, dispelling the notion of a stagnant 10th century.

The impetus for the current book emanates precisely from the interest in Abbo's command of mathematics, coupled with his distinctive philosophical framework. The present work undoubtedly is part of a rich, albeit occasionally intricate, tradition of scholarly investigations explicitly devoted to Abbo. In delineating the state-of-the-art, it is imperative to commence with the year 1935, when André Van De Vyver authored the initial examination of Abbo of Fleury's unpublished works.[4] Subsequently, in the ensuing year, during an address delivered before the *Comité belge*

DOI: 10.4324/9781032643472-1

d'histoire des sciences, Van De Vyver placed Abbo within the stages of early medieval scientific development, hence bolstering the relevance of such a figure in the history of medieval thought.[5]

Despite these pioneering studies, we witnessed a lack of inquiries sufficiently underscoring the significance of Abbo's scientific contributions until the 1980s, when Abbo's mathematical acumen began to attract numerous inquiries.[6] Ron B. Thomson undertook the definitive editing of Abbo's astronomical works, while Barbara Obrist and David Juste's studies brought attention to Abbo's descriptive astronomy, cosmology, and sublunary physics.[7] Furthermore, the scholarly endeavours of Peter Verbist and Nadja Germann have provided additional insights into crucial facets of Abbo's intellectual legacy, with a particular emphasis on the field of computus. Their works elucidate Abbo's methodology in rectifying the chronology established by Dionysius Exiguus (ca. 540 CE), based on Abbo's own luni-solar cycle, while concurrently highlighting the symbolic import of his scientific investigation into the cyclical temporal order.[8] On the other hand, Abbo's musical expertise finally received due acknowledgment once Michel Huglo attributed to Abbo the treatise *Dulce ingenium musicae*, previously edited as anonymous by Michael Bernhard in 1987.[9]

Arithmetic and calculus, positioned by Abbo as preparatory knowledge for his pursuits in astronomy, computus, and music, have become focal points of investigation for scholars including Eva Maria Engelen, Alison Peden, Gillian R. Evans, Charles Burnett, and, with humility, the author of the present book.[10] In elucidating some foundational aspects of arithmetic and calculus, the technical intricacies of these disciplines have been looked at against the backdrop of Abbo's political and theological standings, or even compared to Gerbert of Aurillac's intellectual pursuits.[11] Alison Peden's 2003 edition of the *Commentary on the Calculus of Victorius of Aquitaine* (*Explanatio in Calculo Victorii*), included in the *Auctores Britannici Medii Aevi* series, has heightened interest and also provided the essential groundwork for a systematic exploration of Abbo's thought in its entirety, encompassing not only mathematical but also philosophical and theological intricacies.[12] Subsequent developments, notably owing to the contributions of Anna Somfai and Irene Caiazzo, have significantly reshaped our understanding of the early medieval reception of Platonic philosophy, also through Abbo. This paradigm shift challenges the previously held notion that the first cohesive reflections on Plato's *Timaeus*, the *Commentary* by Calcidius and Macrobius' *Commentary on the Dream of Scipio*, originated with 12th-century thinkers.[13] Such a revised perspective extends to consider the so-called 'physical Platonism' of the 12th-century Chartrian masters as the culmination of a process that commenced precisely at the turn of the first millennium.

This study provides the first systematic overview of Abbo's thought as articulated in his major work, the *Commentary on the Calculus*. Driven by his mathematical interests, Abbo engaged with Victorius of Aquitaine's

Calculus, a compilation of multiplication tables accompanied by a preface elucidating the concept of 'unity'. Originally serving as a valuable resource for computistical purposes, these tables also introduced the use of duodecimal fractions within a calculation framework primarily rooted in the ten-based Roman numeral system. Abbo's *Commentary on the Calculus* emerges as a crucial resource for specialists not only in the history of science but also in the history of philosophy. Abbo's major work unfolds along three main axes: (a) philosophy and theology, (b) arithmetic and calculus, and (c) physics. These axes are thoroughly addressed within this book, following a framework provided in Chapters 1 and 2, which contextualises Abbo and his *Commentary*.

Regarding (a), to which Chapters 3 and 4 are dedicated, an in-depth analysis of Abbo's *Commentary* enables to track the surge of a newly revived philosophical tradition, namely, Neopythagoreanism.[14] On the one hand, Abbo's reading of *Wisdom* 11:21 ("But you have disposed all things by measure and number and weight") leads him to set scientific knowledge within a theological frame, for every creature complies with mathematical criteria that are ultimately of divine origin. On the other hand, he elaborates an original form of 'henology' (i.e., a theory of oneness) that shapes both his account of God and the ontology of all creatures.[15] From this perspective, God is maximally and authentically one, whereas creatures possess a type of unity that can be always fractionated.

Concerning (b), which is the object of Chapters 5 and 6, it is the fractionable unity pertaining to both corporeal and incorporeal entities that is assumed as the object of arithmetic and calculus. Unity and plurality are the two ontological poles enclosing the whole 'fractionable' reality, including the so-called 'artificial compounds'. As such is anything that can be produced following the rules of arithmetic, like, for instance, ratios and proportions among integer numbers. To account for this, the book offers a systematic outline of the basics of medieval speculative and practical arithmetic, showing its implications with philosophy. Therefore, an in-depth reconstruction of some early medieval calculation techniques (e.g., by fingers, by fractions, and by tables) is presented.

Finally, regarding (c), two parallel conceptions are presented in this book – both cohabiting Abbo's thought and both paving the way for further developments in the following centuries. Abbo's physics is tightly linked to quadrivial sciences, for the constitution of the physical realm meets some mathematical, divine paradigms (i.e., number, measure, and weight). In the light of this, the study of the physical realm falls under the branch of philosophy occupied by mathematics. Yet, as explained in Chapter 7, Abbo's physics is not merely 'quantitative' but also 'qualitative' since it is concerned with changes depending on the interaction of qualities that find confirmation in observed natural phenomena and some cases drawn from physiology and metallurgy – binding Abbo's investigation to the tradition of Aristotle's *Problemata physica*.

By reconstructing these three axes, the overall aim of the book is to offer an unprecedented holistic insight into the complexity of Abbo's thought and prompt a re-evaluation of the cultural context in which he lived as well as the science and philosophy of his time.

Notes

1 On this aspect, see Gibson, "The Continuity"; Riché, "Le *quadrivium*" (where the 10th century is referred to as the 'third Carolingian Renaissance'). Cf. Klibansky, *The Continuity*; and Marenbon, *From the Circle*.
2 See, for instance, Materni, "Attività scientifiche"; Mostert, "Gerbert d'Aurillac"; and Otisk, *Arithmetic*.
3 The choice of 'calculus' seems appropriate, given that Abbo's principal work is a commentary on Victorius of Aquitaine's *Calculus*, a compilation of tables featuring arithmetic operations.
4 Van de Vyver, "Les oeuvres inédites".
5 Van de Vyver, "L'évolution scientifique". Abbos is not yet mentioned in an article by Van de Vyver preceding the discovery of his works, see Van de Vyver, "Les étapes du développement".
6 However, during this time, the historiographical landscape was not devoid of activity, as evident in monographs portraying Abbo as a historical figure and uncovering the extant manuscripts and texts coming from Fleury Abbey. See Cousin, *Abbon de Fleury-sur-Loire*; Pellegrin, "*Membra disiecta floriacensa*"; and Pellegrin, "La tradition des textes".
7 Thomson, "Two Astronomical Tractates"; Thomson, "Further Astronomical Material"; Obrist, "Abbon de Fleury"; Juste, "Neither Observation".
8 Verbist, "Abbon de Fleury"; Germann, "À la recherche".
9 Abbo, *Dulce ingenium musicae*, in Huglo, "Le traité de musique"; cf. Abbo, *Dulce ingenium musicae*, in *Anonymi saeculi decimi*. Cf. also Mostert, "Gerbert d'Aurillac"; in this comparative study of Abbo and Gerbert of Aurillac, Mostert referenced a couple of Gerbert's letters addressed to Constantine of Fleury, discussing passages from Boethius' *De institutione musica*.
10 See, for instance, Engelen, *Zeit*; Gillian and Peden, "Natural Sciences"; Burnett, "Abbon de Fleury"; Crialesi, "Un approccio matematizzante".
11 Evans, "*Difficillima et ardua*"; Peden (White), "Unity, Order and Ottonian Kingship"; Peden, "Boethius in the Medieval Quadrivium"; Burnett, "Abbon de Fleury; Crialesi, "Un approccio matematizzante".
12 Abbo of Fleury, *Commentary on the Calculus*. Hereafter, Abbo, *In Calculo*.
13 In this regard, see the statements in Somfai, "The Eleventh-Century Shift", 2: "The mistaken belief that the twelfth century produced the first significant medieval interpretation of Plato's text has so far gone almost completely unchallenged, since little effort has been made to uncover the reception of the *Timaeus* and of Calcidius's *Commentary* from the ninth to the late eleventh century"; and Caiazzo, "Abbon de Fleury", 31: "Abbon de Fleury est sans aucun doute le premier auteur médiéval à avoir une telle familiarité avec les oeuvres de Macrobe et de Calcidius. Il suffira de citer le cas de Guillaume de Conches, l'icône de la renaissance du platonisme du XIIe siècle, auteur des plus amples commentaires conservés sur les *Commentarii in Somnium Scipionis* de Macrobe et le *Timée*, qui est souvent mal à l'aise avec des démonstrations un peu trop techniques touchant à l'arithmétique ou à l'astronomie exposées par Macrobe et par Calcidius, démonstrations devant lesquelles Abbon, lui, ne recule pas". See also the study on the early medieval reception of Macrobius' *Commentary* by White (Peden), *Glosses*.

14 For Neopythagorean fundamental features in Abbo and, more in general, in the history of early medieval philosophy, cf. Albertson, *Mathematical Theologies*, 23–27.

15 On the term 'henology' I refer to the studies by Gilson, *L'être et l'essence*, 42, where the term first made its appearance: "Ce ne sont pas deux ontologies qu'ils s'agit ici de comparer entre elles, mais une 'ontologie' et, si l'on peut dire, une 'énologie'"; Willer, *Henologie*; Beierwaltes, *Denken des Eines*. Cf. also Aertsen, "Ontology and Henology", where Aertsen deals with Gilson's thesis about medieval thinker's effort of reconciling Christian metaphysics with henology.

1 Abbo and His Time

European Culture and Mathematics

1.1 Politics and Religious Culture

For those unfamiliar with the figure and ideas of Abbo – a circumstance anticipated to be prevalent – a brief outline of the historical and cultural context in which he lived is provided to show the profound integration of his profile within the broader European setting.[1] Abbo was born around 945, amidst a century marked by political upheavals yet imbued with profound reformative, political, and cultural impulses. His life is to be situated within an environment that, by contemporary standards, could be described as 'international', in which he actively engaged on both political and intellectual fronts. Two pivotal elements characterise this historical backdrop: the Holy Roman Empire and the monastic reform. Concerning the former, a fresh political stability materialised with the ascendancy of the Ottonian dynasty. Regarding the latter, a restored cultural cohesion emerged from the Cluniac reform.

Abbo's life stands out for his dynamic engagement in contemporary affairs, both secular and religious – a quality already underscored by Pierre Riché's assessment. According to Riché, Abbo is not merely a scholar confined to his monastery's library but rather a man of action and a prominent figure in the political landscape of the 10th century.[2] Throughout his tenure as abbey leader, Abbo dedicated himself to activism aimed at safeguarding the rights of monasteries, a commitment that significantly influenced his theological and political reflections. As mentioned in his *Liber apologeticus*, the scholarly pursuits that once consumed Abbo's time in philosophy yielded to the exigencies of his era, particularly when a dispute arose with Bishop Arnoul of Orléans concerning the episcopal privileges that the monastery of Fleury was expected to adhere to.[3] Abbo, refusing compliance, found himself accused by Arnoul and Gerbert of Aurillac at an episcopal assembly convened at Saint Denis. The charges implicated him in allegedly fomenting a monastic revolt against bishops on the matter of tithes.[4] In direct response to these accusations, Abbo composed the above-mentioned *Liber apologeticus*, directing it towards the Capetian kings. To them he also dedicated the *Collectio canonum*, a compilation of excerpts from canon law texts, which served

DOI: 10.4324/9781032643472-2

as a defence of the monastic order, seeking protection against unwarranted intrusion by both laypersons and bishops.

Furthermore, Abbo embarked on three journeys to the papal see, culminating in the issuance of a bull in 997 by Pope Gregory V (Otto III's cousin). This bull conferred complete autonomy upon the monastery of Fleury. Abbo's active role in the political affairs of his time becomes even more apparent in his acrostic poem *Otto valens*. In this composition, Abbo directs his verses to the Saxon emperor, Otto III, imploring him to cease hostilities with the Slavs. Additionally, he urges the emperor to undertake a journey to Rome in defence of papal authority, which was under threat from John Crescentius. Holding the title of patrician of Rome, John Crescentius controlled the city and had installed an anti-pope, John XVI, on the papal throne.[5]

Abbo's connections with Ottonian and Capetian figures, bishops, and popes intertwine with another pivotal transformation that animated 10th-century society and culture: the monastic reform. This religious reform, emanating from the Cluny monastery, played a crucial role in fostering the revival of educational settings and practices, leading to a transformation of the monastic educational model. Not only did the Cluniac reform influence the structure of monastic life, but it also contributed significantly to the reshaping of educational norms, emphasising a renewed commitment to learning, especially focused on the liberal arts, within the monastic communities.[6] The profound impact of the reform is also discernible in Abbo's perspective on the monastic class. In his *Liber apologeticus*, he situates the monastic order at the pinnacle of a newly defined threefold social structure, consisting of laymen, clerics, and monks.[7]

The influence of the monastic movement extended as far as the British Isles, resulting in the formulation of the *Regularis concordia*, which received official recognition at the Council of Winchester in 973. In England, the restructuring of monasteries gained momentum primarily under the guidance of the Archbishops of York and Canterbury, Oswald and Dunstan, respectively. The adherence to Benedictine practices at Fleury, coupled with the customs detailed by Thierry of Amorbach, served as a blueprint for the reform implemented in numerous English monastic centres – let us remember that Odo of Cluny himself was invited to reform Fleury between 931 and 942. The strong connections between the proponents of the reformation in England and the French abbeys, with Fleury monastery being particularly notable, are widely acknowledged.[8] Oswald, who served as a monk at Fleury from 959 to 961, and Osgar, abbot of Ramsey between 950 and 963, personify these ties.

As the reform solidified the ties between Fleury Abbey and English monasteries, Abbo's engagements extended beyond the borders of the Saxon realm to include travels across the English Channel, notably at Ramsey. As mentioned earlier, monastic schools played a pivotal role in the preservation of a literary culture and the cultivation of interest in liberal arts studies. In reformed English centres, for instance, there is evident effort to revive the use of Latin. A compelling example is the *Quaestiones grammaticales*, which is a Latin manual crafted by Abbo for the monks of Ramsey, wherein Latin grammar blends with references to the local language.[9]

Ramsey is not the only abbey benefiting from Abbo's action. Throughout his life, Abbo was frequently called upon to intervene in numerous tumultuous monasteries, engaging in the reform of abbeys and pacification of internal conflicts. His efforts extended to Marmoutier, Saint-Mesmin de Micy, and Saint-Père en Vallée, where he responded to the invitation of one of his former disciples, Fulbert of Chartres. The final chapter of his interventions took place at La Réole, where he met his demise on November 13, 1004, during a clash between Gascon and Frankish monks. Abbo's sanctity, stemming from his martyrdom, is chronicled by Aimo, monk at Fleury, who intertwines the narrative of Abbo's life with posthumous miracles attributed to him. It was only in 1638 that Abbo was officially incorporated into the martyrology of the Gallican church. According to Aimo's account, just before becoming involved in the tumultuous clash between the two factions, Abbo was engrossed in mathematical calculations within the cloister of the monastery – an image that renders to us Abbo's profound interest in the mathematical sciences, especially arithmetic and calculus.[10]

1.2 Liberal Arts

Outlining Abbo's mastery of liberal arts requires us to first spend a few words regarding the place where he originally initiated his studies: the abbey of Fleury and its library, whose history has been retraced by Elisabeth Pellegrin and Marco Mostert.[11] Established in 651, the Benedictine monastery of Fleury experienced both material and intellectual flourishing following the translation of the relics of St. Benedict in 703, from the abbey of Monte Cassino to Saint-Benoît-sur-Loire (i.e., Fleury abbey). This event marked a significant turning point, as the prestige associated with possessing the remains of the first Benedictine drew pilgrims, resulting in increased donations and concessions from the royal court.[12] Subsequently, Fleury evolved into one of the most distinguished abbeys in the Latin-speaking world, assuming a pivotal role in the cultural and religious milieu of the Carolingian era.

The cultural significance of Fleury is further underscored by the enrichment of its library with manuscripts. According to estimates by Mostert, during the cultural zenith of the monastery, spanning from 800 to 1150, the texts contained within the codicological units of Fleurisian manuscripts numbered around 6000.[13] This extensive collection attests to Fleury's status as a prominent centre of learning and intellectual activity during a critical period in medieval history. From the year 789 onward, the library benefited from the production capabilities of its own scriptorium. While it is likely that the collection at Fleury predominantly featured biblical texts and works by the Church Fathers until the Carolingian era, the state of the library had evolved significantly by the time of Abbo.[14] Two notable strands that emerged were the presence of texts by ancient authors and literature dedicated to logic. Regarding the first strand, information about the amount of ancient literature at Fleury is gleaned from references to classical authors in Abbo's works and other sources, such as Aimo, Abbo's biographer. Notable classical authors mentioned in connection with Fleury include Seneca, Terence, Sallust, Virgil, Horace, Livy, Marius Victorinus, and Cicero.[15] This diversification in the content of the library reflects a broadening intellectual scope and an increased interest in classical literature at Fleury during Abbo's time.

On the subject of logical literature, Anita Guerreau-Jalabert provides insight into manuscripts from Orléans, Bibliothèque municipale, 277, and Leiden, Universiteitsbibliotheek, Voss. lat. F 70, both originating from Fleury.[16] These manuscripts contain foundational texts of ancient logic, including Porphyry's *Ysagoge* with Boethius' commentary, Aristotle's *Categoriae* and *De interpretatione* in translation with commentary by Boethius, Cicero's *Topica*, and Apuleius' *Peri hermeneias*. Additionally, chapters on dialectics by Isidore of Seville, Martianus Capella, and Alcuin, along with the *Categoriae decem* of Pseudo-Augustine, are found to be in connection with manuscripts stemming from Fleury library. Notably, Fleury's collection also includes Boethian treatises on categorical and hypothetical syllogisms and topics, showcasing the uniqueness of its intellectual holdings. The origin of these Boethian works can be traced back to the Byzantine Empire between 522 and 526, when they were collected by Martius Novatus Renatus and copied in Constantinople by Theodore, a disciple of Priscian. Their reappearance in Europe did not occur until the mid-10th century, specifically in northern France. Fleury's library, therefore, seems to have played a crucial role in reintroducing these significant logical texts to Western monastic intellectual circles.[17]

The educational system at Fleury, where Abbo received his training and later actively participated as a teacher, is elucidated by Thierry of Amorbach in the *Consuetudines Floriacenses*. The organisational structure of intellectual pursuits at Fleury centred around the pivotal role of the *armarius*. At Fleury, the *armarius* played a multifaceted role, overseeing various aspects of the intellectual life within the monastery. This included the meticulous care of the library's resources, responsibility for activities associated with manuscript copying (from procuring materials for codices to correcting copied texts), and simultaneous dedication to the education of students.[18] To comprehend the nature of monastic education available at Fleury, one can also turn to a concise note written by Abbo's most famous disciple, Byrhtferth of Ramsey. In the concluding section of his *Enchiridion*, Byrhtferth delves into a digression on the symbolism of numbers (*De numerorum significationibus*), incorporating a table for the number 1000.[19] It is with regard to this number that Byrhtferth recalls his master, emphasising two significant aspects: the dignity of Abbo's life and his expertise in mathematics, culminating in perfection in philosophy.[20] These features are consistently highlighted in a passage from Aimo's *De vita et martyrio Sancti Abbonis*, where it is explicitly mentioned that Abbo's monastic life, characterised by unwavering dedication, extended beyond mere prayer practices. Aimo notes an additional facet to Abbo's activities – a form of relief for Abbo himself – centred precisely around the pursuit of the liberal arts.[21] Liberal arts appear to be the core of Abbo's expertise and, thus, the core of the education provided at Fleury.

Born around 945 into a modest family in the Orléans region, Abbo's early education commenced when he was introduced as an oblate to the abbey of Saint-Benoît-sur-Loire in Fleury at the age of seven. This move was made with the specific aim of providing him with a generic level of learning with a grammatical foundation ("litteris imbuendus", "profunda litterariae artis", "litterarum exercitium").[22] It was only in his early youth that Abbo was initiated to a more complex kind of learning, which also encompassed chant, liturgy, and the study of texts dedicated to some liberal arts, especially grammar, logic, and arithmetic. At the end of this educational track, Abbo eventually assumed the role of lecturing Fleury's students, taking charge of their education ("imbuendis preficitur scholaris").[23]

The pedagogical scope at Fleury was relatively limited, as it was in other monasteries of the time, primarily focusing on elementary education in the grammatical sphere. Only a handful of monastic centres could offer teaching in less conventional disciplines. Consequently, during Abbo's era, intellectuals often sought out those centres where the presence of a distinguished master was guaranteed, ensuring specialised education in a specific discipline.[24] Indeed, Abbo himself experienced the desire to broaden his liberal knowledge beyond grammar, logic, and arithmetic. Faced with this aspiration, he embarked on a journey, reflecting the common practice of learned men during that time.[25]

In the second half of the 10th century, Abbo sought to complete his training by venturing to Paris, specifically to Saint-Germain-des-Prés. It is likely that there were still disciples of Remigius of Auxerre in Paris during that time. Remigius had relocated to Paris from Reims, a city that Abbo himself would later visit. In Reims, Remigius, at the invitation of the archbishop Fulfus, had instructed students in grammar and logic, introducing them to the works of Martianus Capella and John Scotus Eriugena. During Abbo's visit to Reims, Adalberon, the bishop of Reims at that time, aimed to provide liberal arts instruction to the students of his church. To achieve this, he enlisted the help of the esteemed archdeacon Geranus, known for his reputation as a logician. It is likely that Abbo sought guidance from Geranus during this period. Three years later, Gerbert of Aurillac also studied under Geranus. As suggested by Elizabeth Dachowsky, it is improbable that Abbo met Gerbert during this time, for if Gerbert and Abbo were in Reims simultaneously, Abbo would have likely pursued geometry, too, a discipline in which Gerbert was one of the few experts in all of Frankish kingdom.[26] According to Aimo, however, during his time in Paris and Reims, Abbo focused on advancing his knowledge especially in the field of astronomy.

Orléans marked the final leg of Abbo's educational journey. During his time in Orléans, he sought music theory lessons from a cleric, paying a fee for these lessons in secret. This is also reflected in the prologue of Abbo's *Explanatio*, written a few years after his stay in Orléans. There, Abbo articulates a critique of the venality of certain teachers of liberal arts – Abbo was critical of the practice of some educators who charged fees for their instruction in the liberal arts, emphasising the importance of education as a noble pursuit rather than a commercial venture.[27]

After returning to Fleury and having attained proficiency in most of the liberal arts (grammar, logic, arithmetic, music, and astronomy), Abbo continued his intellectual pursuits. He delved into the subjects of geometry and rhetoric as a self-taught scholar, making his expertise unparalleled among his contemporaries. Simultaneously, Abbo dedicated himself to teaching at Fleury and authored some of his early works. These works included a treatise on syllogistics (Aimo states: "dialecticorum nodos syllogismorum enucleatissime enodavit"), an introduction to practical arithmetic (i.e., the *Explanatio*), calculation tables for computes ("in more tabularum texuit calculationes"), and several astronomical treatises ("de solis quoque ac lune seu planetarum cursu").[28]

Between 985 and 988, Abbo journeyed to Ramsey, a significant centre of monastic reform in England. His visit was prompted by an invitation extended by an English delegation that had travelled to Fleury in search of a master. During his time at Ramsey, Abbo not only contributed to the biographical account of St. Edmund but also authored a work on Latin grammar (i.e., the *Quaestiones grammaticales*) and completed other works on astronomy. This period of Abbo's life showcases his role in disseminating knowledge and contributing to the intellectual endeavours of monastic communities beyond the confines of Francia.

Zooming into Abbo's works, one can easily divide them into those dedicated to the disciplines of the trivium (specifically grammar and logic) and those focusing on the quadrivium or mathematics (specifically astronomy, computus, music, calculus, and arithmetic). During his teaching tenure at Fleury, following his study visits, Abbo composed two significant works: *De syllogismis hypoteticis* and *De syllogismis categoricis*.[29] While the latter adheres to the conventional systematisation of syllogistic treatises, the former is grounded in a Boethian text known precisely as *De syllogismis hypoteticis*. This Boethian text had been unknown in the Latin West until the 10th century. The presence of Boethius' *De syllogismis hypoteticis* in the region of Reims and Saint-Benoît-sur-Loire is believed to have been facilitated by a gift from the Byzantine emperor to the king of Francia. According to Franz Schupp's suggestion, the king likely sent the text to the Fleury scriptorium for copying. Abbo, thus, emerges as a scholar of the 10th century who had the unique opportunity to engage with and work on this text, making him a notable figure in the study of Boethian logic in that period.[30] The *Quaestiones grammaticales*, in contrast, stand out as a pedagogical work meant for the brethren of Ramsey. This linguistic treatise, combining descriptive and normative elements, is designed to facilitate the learning of a language that had become unfamiliar to the inhabitants of the region. The exploration of phonetic, morphological, and syntactic domains within the *Quaestiones* paints a portrait of a Latin language that blends classical and late antique models but is fundamentally rooted in Carolingian grammatical traditions. The work serves as a valuable resource for understanding the linguistic dynamics and educational practices related to Latin in the context of Ramsey at the turn of the first millennium.[31]

In both Fleury and Ramsey, Abbo engages in the writing of astronomical and computistical works, although it is worth noting that these works are comprised of relatively small textual units. Those pertaining to astronomy are compiled under the titles *Sententia Abbonis de differentia circuli et spere*, *De duplici ortu signorum*, and *De quinque circulis mundi*. Additionally, Abbo composes an acrostic poem focused on Higynus' *Astronomia* and titled *In patris natique simul*.[32] Around 978, Abbo initiated the revision of Helperic of Auxerre's *Liber de computo*. This revision served as a prelude to the creation of Abbo's original work, the *Computus vulgaris* or *Ephemerida*, which took the form of an astronomical poem.[33] From his return from Ramsey until the end of his life, Abbo dedicated himself to the reform of the Dionysian calendar. He completed his own computus with a new set of lunar tables and a luni-solar cycle, accompanied by a *Praefatio* introducing paschal cycles. In addition to these contributions, Abbo produced the *Prologus de ciclo magno paschae*, two more cycles 'secundum ordinem annorum', and a luni-solar table of the 'recurrent days' known as the *Laterculus anterior*.[34] In letters dating back to 1003 and 1004, addressed to Gerald and Vitale, Abbo delves into the cycles 'secundum fidem historiographorum' and presents a revised version of the luni-solar table of the 'recurrents', referred to as the *Laterculus posterior*.[35]

The *Dulce ingenium musice* likely originated from the knowledge Abbo acquired during his studies in Orléans. This work, commencing with a practical examination of the division of the monochord, includes explicit references to Boethius' *De institutione musica*, particularly in relation to the Greek modes, and to Book IX of Martianus Capella's *De nuptiis*. It provides an exposition on *symphoniae*, encompassing simple and compound consonances, along with a commentary on the structure of the eight tones of the Gregorian chant, presented in the *Nova expositio*.[36]

Moreover, the so-called 'Abbo's abacus' stands as an embodiment of Abbo's mathematical prowess.[37] This numerical table, which is not actually an abacus, configured as a diptych with six columns and nine lines, carrying the title *In hac figura descriptus est numerus infinitus*. It is accompanied by some multiplication rules attributed to Abbo. Notably, the same table is also present in Byrhtferth's *Enchiridion*, particularly in the section discussing the number 1000.[38]

In addition to the aforementioned mathematical works, one must include the *Explanatio in Calculo Victorii*, which is the subject of this book. The *Explanatio* serves as an introduction to speculative arithmetic and calculus, covering both integer numbers and fractions. It not only showcases Abbo's mastery of the technical aspects of arithmetic but also stands out as an example of his proficiency in philosophical matters. This work highlights Abbo's ability to bridge the realms of mathematics and philosophy – and theology, too, as it will be shown. Abbo's mastery of arithmetic reflects his position in the continuum of mathematical knowledge, with each era contributing to the cumulative understanding and refinement of arithmetical concepts and practices.

1.3 Mathematics

Abbo's mastery of arithmetic places him within a rich lineage of mathematical development across the centuries. Arithmetic, as a branch of mathematics, has evolved significantly over time, with contributions from numerous scholars shaping its progress. A brief overview of what happened in this field of knowledge – from the early to the late Middle Ages – will help us contextualise Abbo's place in the history of mathematics as well as appreciate his contribution to both speculative and practical arithmetic.

1.3.1 Speculative Arithmetic

In the context of arithmetic during the Middle Ages, it is essential to recognise the diverse manifestations of this discipline, shaped by varying methodologies, purposes, and objects of study. A crucial distinction can be drawn between speculative and practical arithmetic. Speculative arithmetic, encompassing what is often referred to as number theory, involves the examination of numbers in themselves, treating them as discrete, unrelated quantities not necessarily tied to external factors like, for instance, motion. This branch delves into the essential properties of numbers (such as even and odd) and the relationships between them. Its focus, thus, is not on numbers as subjects of arithmetical operations or calculus but on their nature. During the chronological period under consideration, this form of arithmetic primarily dealt with natural numbers, that is, positive integers. Prominent figures in the development of speculative arithmetic include Martianus Capella (360–ca. 428), Boethius (480–524), Cassiodorus (ca. 487–ca. 583), and Isidore of Seville (560–636). In later centuries, scholars like Jordanus de Nemore (*fl.* 13th century) and Thomas Bradwardine (ca. 1290–1349) continued to contribute to this type of arithmetic.

One specific text proves to be of paramount importance for the evolution and subsequent discussions surrounding arithmetic throughout the Middle Ages and beyond: Boethius' *De arithmetica*.[39] This is a fundamental work not only in the domain of number theory but also in the philosophy of arithmetic itself, for it encompasses philosophical examinations on the nature and status of numbers. Many of the reflections found in Boethius' work resurface in later arithmetical treatises – both speculative and practical. Let us have a look at its contents and argumentative structure, and subsequently compare it with another influential text on medieval speculative arithmetic, namely Jordanus' *De elementis arithmetice artis*.

Written around 500, the *De arithmetica* serves as a rather liberal paraphrase of Nicomachus of Gerasa's *Introduction to Arithmetic* (Ἀριθμητικὴ εἰσαγωγή), a Neopythagorean work from the 2nd century AD.[40] Given its roots in Neopythagorean thought, the reflection on arithmetic and, more significantly, on the concept of number presented by Boethius aligns neatly with the Neopythagorean model.[41] The *De arithmetica* is a systematic treatise, comprising two books that contain 32 and 54 chapters, respectively. However, this division into books and chapters may not provide optimal clarity for the reader. To facilitate an understanding of its content, it is more suitable to thematically divide the text into five sections. The first section is particularly relevant from a theoretical standpoint, focusing on the philosophy of arithmetic and numbers – I will come back to it shortly.[42] A succinct synopsis of the content in the other sections will further elucidate the nature of number theory. In the second section, Boethius introduces a classification of numbers based on their even or odd nature. This classification further extends into sub-categories such as even-times even, even-times odd, and odd-times even. Additionally, Boethius delves into the categorisation of numbers as perfect, imperfect, and superabundant, all grounded in the divisibility of numbers and the identification of their factors.[43] For instance, numbers like 6 and 28 are classified as perfect because they equal the sum of their own factors (1+2+3=6; and 1+2+4+7+14=28).[44] The third section of the text focuses on ratios (*proportiones*) between whole numbers, with Boethius identifying five distinct types.[45] In modern mathematical terminology, only the term 'multiple' remains, representing a specific type of ratio involving two unequal quantities where the greater quantity contains the lesser one an exact number of times.[46] In the fourth section, Boethius shifts the focus to the various ways in which numbers can be described and classified based on their geometrical representation. These are referred to as 'geometric numbers'. The classification includes numbers that can be represented through certain plane and solid geometric figures, such as linear numbers, squares, cubes, pyramidal numbers, spherical numbers, and more.[47] The final section delves into different types of proportions, with Boethius primarily presenting three: the arithmetical, the geometrical, and the harmonic.[48]

Six centuries after Boethius, Jordanus de Nemore composes the other great text of speculative arithmetic: the *De elementis arithmetice artis*, which gathers arithmetical notions from Boethius' *De arithmetica* and Book VII of Euclid's *Elements*. A concise overview of the contents of the *De elementis arithmetice artis* reveals his theoretic dependence precisely from Boethius' and Euclid's arithmetic: analysis of what is unit, number, and part (Book I);

properties of equality and inequality ratios (Book II); prime and composite numbers (Book III); 'proportional' numbers, including multiple and submultiple (Book IV); multiplication and division of ratios (Book V); plane and solid numbers, e.g., square, rectangular, cubic, etc. (Book VI); even and odd numbers (Book VII); 'geometric' numbers (Book VIII); five species of inequality ratio (Book IX); and proportions (Book X). The relevance of the *De elementis arithmetice artis*, thus, is not to be found in the originality of its contents, but rather in the specific way they are organised and presented. As evident from the title of the work, Jordanus' intention is to provide the arithmetic equivalent of Euclid's *Elements*, aiming to offer the 'elements', that is a foundation of arithmetic: Jordanus' aim to deliver an axiomatic treatment of this disciple is evident from the organisation and exposition of the contents. Boethian and Euclidean arithmetic are rearranged according to definitions, axioms, and postulates, organised according to propositions, each accompanied by the corresponding proof (in which letters of the alphabet are used instead of numbers). This axiomatic structure is reflected in the division of propositions into definitions (*definiciones*), petitions (*peticiones*), and common conceptions of the soul (*communes animi conceptiones*), a structure similar to the Latin versions of Euclid's *Elements*, with the labels of the parts of propositions being closer to the version known as Adelard II, likely authored by Robert of Chester.[49]

Evidently, the type of exposition and argumentation employed by Jordanus is not the same utilised by Boethius. This comparison can be articulated by paraphrasing the words of Roger Bacon. In his *Communia mathematica*, when Bacon introduces speculative arithmetic, he mentions that this discipline deals with the species, passions, attributes of numbers, and with the causes from which we derive what is useful in the sciences and in the understanding of material things. Euclid and Boethius, according to Bacon, present this through a *propter quod* explanation. Still in Bacon's view, although Boethius does not proceed with demonstrations in the same manner as Jordanus, the cogency of Jordanus' demonstrations somewhat derives from Boethius. Therefore, Boethius' *De arithmetica* is considered an excellent and more useful book than Jordanus', as many things to be established about numbers do not actually require Jordanus' chosen mode of demonstration.[50] Despite Bacon's sharp judgment, who nonetheless relies on Jordanus for his *Communia mathematica*, it is essential to note Jordanus' association with an axiomatic type of exposition, a characteristic that does not pertain to Boethius' treatise. Indeed, the *De elementis arithmetice artis* represents a reconsideration of speculative arithmetic in axiomatic terms, marking a pivotal moment in the history of this discipline.

Another notable distinction between Jordanus' text and Boethius' is the absence of the 'philosophical section' and its philosophical implications in Jordanus' work. It is worth discussing this section of Boethius' *De arithmetica*, also to show how Boethius' Neopythagoreanism emerges. Such Neopythagoreanism can be summarised in three points. (1) From an epistemological perspective, the study of arithmetic, and mathematics in general, is considered the initial step towards wisdom, which involves knowledge of essences. Initiating the journey to wisdom must involve delving into the essence of quantity; for if one dismisses mathematics, then they cannot achieve any philosophical knowledge.[51] (2) Another Neopythagorean aspect pertains to the ontology of numbers. Numbers, positioned between God and creatures, are not subject to change.[52] (3) Moreover, metaphysically, numbers serve as the primary paradigm or model through which God structures the entire universe. They establish metaphysical conditions governing the unfolding and succession of physical phenomena, ensuring harmony in creation.[53]

Although the reflections on Neopythagorean elements are absent in Jordanus' *De elementis arithmetice artis*, they resurface unexpectedly in other works that, at first glance, may appear unrelated to speculative arithmetic but are instead representative of practical arithmetic. Notably, these Neopythagorean elements find expression in the work of John of Holywood. In his *Algorismus vulgaris* (or *De arte numerandi*), a treatise on practical arithmetic widely circulated for its use as a university text, John of Holywood immediately incorporates a quotation from Boethius' *De arithmetica*, specifically espousing the above-mentioned feature (3).[54] In John's view, the study of practical arithmetic gains significance in light of a broader metaphysical assertion that posits the formation of everything in accordance with numerical rationality. This overarching claim grounds the understanding the science of calculus and algorisms.

In this theoretical spectrum, with Boethius, on the one hand, and Jordanus de Nemore and John of Holywood, on the other, as the endpoints, Abbo positions himself not exactly in the middle: if it is true that he leans more towards a Boethian approach to the treatment of speculative arithmetic, as any axiomatic reorganisation of the discipline of arithmetic is absent in the *Explanatio*, it is also true that Abbo incorporates all the essential elements of speculative arithmetic (e.g., inequality ratios, perfect numbers, geometric numbers) to constitute the frame where other expressions of practical arithmetic gain significance – they will be detailed in the next section. The remarkable aspect of Abbo's work lies precisely in the ability to seamlessly integrate both speculative and practical arithmetic, that is, to combine number theory and calculus. Similar to John of Holywood, Abbo, too, sees the combination of speculative and practical arithmetic as imbuing calculus with philosophical value, rooted particularly in Neopythagorean assumptions, notably points (1) and (3).

1.3.2 *Practical Arithmetic*

Practical arithmetic can be categorised based on the methods and tools it employs to perform primary calculations. I call this branch of arithmetic 'calculus', encompassing arithmetical operations with integers and fractions carried out with various methods. One form of practical arithmetic involves the use of hands (and other body parts) to implement said computations. This method is known as *loquela digitorum* (literarily: the speech of fingers), where specific values are assigned to the digits of multiplicands and products based on their positions on various joints and fingers of the hand. An example of this type of arithmetic is presented in the 8th century by Bede the Venerable in his work *De temporum ratione*. References to this method, including not only joints, fingers, and hands, but also other parts of the body, can be traced as far back as Martianus Capella's Book VII of the *De nuptiis Philologiae et Mercurii*. There, the character of Arithmetic states that she accepts numbers that can be represented only by fingers rather than by a 'twisted dance with the arms'.[55] In the 10th century, Abbo will include such a way of calculating in his *Explanatio*, detailing the value to be assigned to joints and fingers.[56]

Another manifestation of practical arithmetic is found in abacus treatises. In these works (which oftentimes consist only of tables accompanied by concise rules), the focus is on the computational operations that can be performed using specific grids of columns (usually 27 and grouped in threes by arches) that have specific values.[57] A notable advancement in this form of arithmetic is the introduction, albeit partial and gradual, of the positional system. In the early manifestations of the abacus, the emphasis is not so much on the notation but on the place-value method that assigns a power of ten to each column, in which Roman numerals (or later on, Arabic-Indian numerals) can be posited. The positional representation, in this case, involves assigning a specific value to a given column, such as thousands, hundreds, tens, and so on. Some treatises on the abacus, precisely those by Gerbert of Aurillac (945–1003), began incorporating Arabic-Indian numerals – or at least the 'Western form' of these numbers.[58] About this specific notation, let us know that during the 11th and 12th centuries, medieval Europe shared a common mathematical language, with Greek, Latin, and Arabic masters referring to the Arabic-Indian numeral symbols as 'Indians' due to their place of origin.[59] However, there were variations in the adoption of certain Indian numerals by Latin masters, leading to the distinction between 'Eastern numbers' and the 'Western numbers' – the latter being closer to the one still in use today in Europe. From Gerbert of Aurillac's utilisation of Western numerals in his abacus, it is likely that the so-called *apices*, that is, Latin pebbles or counters, are derived.[60] In the early Middle Ages, authors who dedicated themselves to this type of arithmetic, as previously mentioned, include Gerbert of Aurillac, Abbo of Fleury, Bernelinus (*fl.* 1000), Hermann of Reichenau (1013–1054), Turchillus (*fl.* first half of 12th century), Ralph of Laon (d. 1131), and Adelard of Bath (ca. 1080–1151).

Arithmetical operations (or what I refer to as calculus) in the early Middle Ages extended beyond integer numbers to include fractions (i.e., *unciae* and *minutiae*). Fractions were conceived as aliquot parts of a whole, known as *assis*: such aliquot parts were conceived as divisions of a whole that can be fractioned on base 12. Moreover, they were expressed using specific symbols rather than the numerator-denominator form familiar today. This notation persisted until Leonardo Fibonacci introduced the horizontal bar in fractions, later popularised by John of Lignères in his *Algorismus de minutiis*. Despite this evolution, early medieval notation and names for fractions persisted alongside the newer representation still in Fibonacci's *Liber abaci* and beyond.[61] Duodecimal fractions, originally introduced in antiquity for measuring weight and time, retained significance in medieval Europe – some medieval fraction names, such as the 'ounce', are still used in weight measurement today. Abacus and multiplication tables, including those developed by Victorius of Aquitaine in his *Calculus* (i.e., the tables commented on by Abbo), incorporated fractions, especially as a prelude to time reckoning.[62] Abbo's contributions to the calculus with fractions stand out as foundational, encompassing a systematic overview of their symbols, names, and relationships. He integrated fractions into tables of diverse arithmetical operations and discussed equivalence tables linking fractions to weight and time measures. Abbo, thus, played a pivotal role in consolidating diverse facets of fractions during his time.[63]

During the 13th century, a shift occurred in the realm of practical arithmetic with the rise of treatises known as 'algorisms', marking a transition from earlier methods. This evolution is exemplified in Leonardo Fibonacci's *Liber abaci*, which serves as a bridge connecting traditional practical arithmetic to the emerging algorisms, and by Jordanus de Nemore's *De numeris datis*. The term 'algorism' derives from the Arabic mathematician Al-Kwarizmi, known for his treatises on arithmetic and algebra. His works, including the *Treatise on Calculation with Hindu Numerals* and the *Book on Union and Separation*, were translated into Latin, titled *Algoritmi de numero indorum* (or *Libri alchorismi de practica arismetice*) and *Liber augmenti et diminutionis*. This Latinised nomenclature influenced subsequent Latin authors. Arithmetic, more and more evolving into the science of calculus (*ars numerandi*), focused on seven primary operations: addition, subtraction, duplication, halving, multiplication, division, and root extraction. These operations utilised the Western Arabic-Indian notation within a decimal positional system. In the 13th century, European authors produced treatises influenced by the *De numero indorum*, contributing to the proliferation of works on algorism. Among the notable works were the *Carmen de algorismo* by Alexander of Villedieu (1175–ca. 1240) and the above-mentioned *Algorismus vulgaris* by John of Holywood (ca. 1195–ca. 1256). These works became integral components of liberal arts programmes at various universities. It is possible to identify fundamental differences between these two methods of performing arithmetical operations, namely, abacus and algorism. For instance, abacus authors list names and numerals for both integers and fractions, while algorists present only Arabic figures. Abacus authors, though leaving a blank spot in a column to represent zero, did not seem to fully utilise the corresponding symbol, which was instead employed by algorists. Abacus treatises show limitations in their range of operations, while algorisms include various 'species' of numerations, such as the extraction of roots, doubling, and halving. Regarding fractions, abacus masters adhered to a base 12, while the decimal base became the norm for algorists.[64] It is worth specifying that algorisms often overlapped with the first developments of algebra. Moreover, the abacus I deal with in this section should not be confused with the tradition of 13th- and 14th-century Italian 'scuole d'abbaco', which provided mathematical education that extended to algebra and was directed towards commercial and recreational goals. For chronological reasons, Abbo does not trace back to the scientific tradition associated with algorisms nor to that related to Italian abacus schools, marking a limit to our historical knowledge of his contributions in this particular context.

Notes

1 Portraying Abbo as a European figure aligns with the approach taken by scholars in examining his more renowned contemporary, Gerbert of Aurillac, with whom Abbo shares historical prominence. See, for instance, Charbonnel and Iung (eds.), *Gerbert l'européen*.
2 Riché, *Abbon de Fleury*, 125.
3 Abbo of Fleury, *Liber apologeticus*, 461 B-C.
4 Events are more thoroughly recounted in Riché, *Abbon de Fleury*, 140–146.
5 Abbo's acrostic poems have been studied by Gwara, "Three Acrostic Poems", specifically for the *Otto valens*, see 228–231.
6 On liberal studies within monastic contexts, see Lesieur, *Science des nombres*. On mathematical disciplines like music and astronomy in particular, see McCarthy, *Music*; A. Borst, *Astrolab*; and Borrelli, *Aspects of the Astrolabe*.
7 On Abbo's political views, see Mostert, *The Political Theology*; Batany, "Abbon de Fleury".
8 Connections between Fleury Abbey and the English monastic centres have been investigated by Gougaud, *Les relations*. For Thierry of Amorbach, I refer to Davril, "Un moine de Fleury".
9 Riché, *L'enseignement*, 25–26; and Abbo of Fleury, *Quaestiones grammaticales*.
10 Aimo, *Vita Sancti Abbonis*, 20, p. 123: "Intra claustrum monasterii residens et quasdam compoti ratiuncula dictitans".
11 Pellegrin, "Membra disiecta Floriacensa"; Pellegrin, "La tradition"; Mostert, *The Library*.
12 Leclercq, *Saint-Benoît-sur-Loire*.
13 Mostert, "La bibliothèque".
14 Mostert, *The Political Theology*, 32–33. On the basis of the presence of Greek manuscript at Fleury, Riché put forth the hypothesis that Abbo might have known the Greek. See Riché, *Abbon de Fleury*, 83–84.
15 These authors are mentioned in Abbo's *Quaestiones grammaticales*, *In Calculo*, and astronomical works. See also Aimo, *Vita Sancti Abbonis*, 3, p. 49. About Cicero, one of Gerbert of Aurillac's letters stands out as particularly significant. In 986, Gerbert requested the monk Constantine of Fleury to bring Cicero's *De re publica* to Reims. During medieval times, only an excerpt from Book VI, known as the *Somnium Scipionis*, circulated. Although it is inferred that this text reached Fleury, concrete evidence is lacking. In the 18th century, Dom Gérou reported that a copy of Cicero's *De re publica*, preserved in the library at Orléans and in all likelihood coming from Fleury abbey, was burned in 1562. See Gerbertus,

Epistulae, 86, p. 204: "Comitentur iter tuum Tulliana opuscula, vel de Re publica, vel in Verrem, vel quae pro defensione multorum plurima Romanae aeloquentiae parens conscripsit". Cf. Mostert, *The Library*, 27 and 249 (BF1291).

16 See Guerreau-Jalabert's "Introduction" in Abbo, *Quaestiones grammaticales*, 190–191.

17 Riché, *Abbon de Fleury*, 98–99. Van de Vyver, "Les étapes" 443–446. See also Troncarelli, "«Haecine est bibliotheca ...»", 21–23; and Casey, "Boethius's Works", 212.

18 Cf. the description of Fleury's *armarius* in Thierry d'Amorbach, *Consuetudines Floriacenses*, 9, pp. 182–184. Cf. Mostert, *La bibliothèque*, 120.

19 The table coincides with the one found in the codex Oxford, St. John's College 17, fol. 35v. We will come back to this table later in Chapter 6 of the present book.

20 Byrhtferth, *Enchiridion*, IV, 1, p. 228: "Ratio huius numeri, quam digne memorie Abbo super hunc invexit, libet libari. Iste vero quante dignitatis refulsit in vita ostendunt post mortem miracula. Erat enim in doctrinali scientia peritus et in philosophia perfectus". Since Cassiodorus, mathematics had been defined as *doctrinalis scientia*; see Cassiodorus, *Institutiones* II, 3, 21 p. 130: "Mathematica, quam Latine possumus dicere 'doctrinalem', scientia est quae abstractam considerat quantitatem"; cf. Isidore of Seville, *Etymologiae*, III, p. 3, 21, 109: "Latine dicitur doctrinalis scientia, quae abstractam considerat quantitatem. Abstracta enim quantitas est, quam intellectu a materia separantes vel ab aliis accidentibus, ut est par, inpar, vel ab aliis huiuscemodi in sola ratiocinatione tractamus. Cuius species sunt quattuor: id est Arithmetica, Musica, Geometria et Astronomia".

21 Aimo, *Vita Abbonis*, 2, p. 47: "Religionis namque propositum monastice quod habitu pretendebat, id tota mentis devotione diligebat et quasi pro quodam levamine post precum ad Deum missa libamina liberalium artium sumebantur exercitia".

22 Aimo, *Vita Abbonis*, 2, p. 42. Cf. also Guerreau-Jalabert, "Introduction", in Abbo, *Quaestiones grammaticales*, 18–19. The study by Dachowsky, *First among Abbots* allows for filling the gaps left by Aimo and for diligently retracing Abbo's career.

23 Riché, *Abbon de Fleury*, 26.

24 See Riché, *Écoles et enseignement*, 137–147; Riché and Verger, *Des nains*, 59–63.

25 The detailed recollection of this journey is in Aimo, *Vita Abbonis*, 3, p. 49.

26 Dachowski, *First among Abbots*, 46.

27 *In Calculo*, I, 2, p. 64.

28 Aimo, *Vita Abbonis*, 3, p. 49.

29 Abbo of Fleury, *Syllogismorum categoricorum et hypotheticorum enodatio*; Abbo of Fleury, *De hypoteticis syllogismis*.

30 Schupp, "Abbon de Fleury", 48–49 for the codex Renatus (Paris, BnF, nouv. acq. lat. 1611) preserving Boethius' logical works.

31 Cf. Guerreau-Jalabert, "Introduction", in Abbo, *Quaestiones grammaticales*, 83–85.

32 Abbo, *In patris natique simul*, 24–27.

33 Baker and Lapidge, "More Acrostic Verse", 12.

34 Verbist, "Abbo of Fleury". The "recurrent days" (*dies concurrentes*) is a technical notion in medieval time reckoning. Generally speaking, it refers to the number of days that needs to be added to a cycle (e.g., the lunar one) to comply with the duration of another cycle (e.g., the solar one). As explained by Immo Warntjies, since Dionysius Exiguus, *concurrentes* "can be defined as the weekday (*feria*) of a fixed Julian calendar date, and in the early Middle Ages, it was traditionally understood to refer to 24 March. Since the weekday data is dependent on the 28-year solar cycle, the *concurrentes* recur in exactly the same order every 28 years". See Warntjies, *The Munich Computus*, 343.

35 Cordoliani, "Abbon de Fleury", especially 476–480.

36 Huglo, "Le traité". For the intricacies of early medieval music theory, see Huglo, "Gerbert, théoricien".
37 The table will be discussed in Chapter 6 and referred to as 'Ramsey table' (a table transmitted in manuscript Oxford, St. John's College, 17).
38 It is precisely in this passage that Byrhtferth introduces the figure of Abbo as a master having obtained a perfect expertise in mathematics and philosophy. Byrhtferth, *Enchiridion*, IV, 1, p. 228.
39 The bibliography on Boethius' *De arithmetica* is massive. Contributions that help frame Boethius' mathematical thought and provide further bibliographic references are: Masi, "The Influence"; Gibson, *Boethius*; Pizzani, "Il *quadrivium* boeziano"; Guillaumin, "Boèce traducteur"; Guillaumin, "Le status des mathématiques"; Guillaumin, *Boethius's* De institutione arithmetica; and Crialesi, "The Status of Mathematics".
40 Cassiodorus' and Isidore's reflections on arithmetic stem from Nicomachus, too. Nicomachus' *Introduction to Arithmetic* was transmitted to the Latin-speaking world not only via Boethius' translation, for according to Cassiodorus, also Apuleius (125–170) translated Nicomachus' text into Latin. See Cassiodorus, *Institutiones*, II, 4, 7, p. 140: "Quam apud Graecos Nicomachus diligenter exposuit. Hunc prius Madaurensis Apuleius, deinde magnificus vir Boethius Latino sermone translatum Romanis contulit lectitandum".
41 On the Neopythagorean and Aristotelian epistemic models in Boethius' thought, see Crialesi, "The Status of Mathematics".
42 Boethius, *De Arithmetica*, I, 1–2, pp. 6–12. Henceforth *Arithm.*
43 *Arithm.*, I, 3–20, pp. 12–45.
44 Most of these topics are expounded in Abbo's *Explanatio* as well (*In Calculo*, III, 17, pp. 82–83). For the symbolic value of perfect number in Abbo's account of compounds, see Chapter 5.
45 Abbo thoroughly addresses all inequality ratios (*In Calculo*, III, 22–27, pp. 85–89 and III, 72–83, pp. 116–123), which are explained in Chapter 6 of the present volume. The so-called 'rule of Adrastus', as outlined in *Arithm*, II, 1, pertains to a mathematical procedure through which all five types of inequality ratios can be derived from a triplet of numbers in the equality ratio of 1:1. Additionally, this rule serves to revert the numbers back to the equality ratio, namely 1:1. Such a rule is detailed in Chapter 6.
46 *Arithm.*, I, 21-II, 3, pp. 45–87.
47 *Arithm.*, II, 4–39, pp. 88–139.
48 *Arithm.*, II, 40–54, pp. 140–177.
49 Cf., for instance, Jordanus de Nemore, *De elementis arithmetice artis*, I, pp. 64–65, and *Robert of Chester's (?)*, 113–115.
50 Roger Bacon, *Communia mathematica*, I, 4, p. 47: "Deinde sequitur Arismetica cuius speculativa docet species numerorum et passiones eorum et causas absque eo quod descendat ad opera utilia in scienciis et rebus, et hec traditur apud Euclidem et Boetium per viam cause et propter quod, et apud Marciannum et multos per modum narracionis et quia est. Boecius autem, licet secundum modum procedendi non videatur sic demonstracionibus uti sicut Jordanus, tamen virtus et sciencia demonstracionum Jordani ex fontibus Boetii oriuntur. Unde optimus liber est *Arismetica* Boetii et utilior quam liber Jordani, quia pauca de numeris demonstranda sunt sub modo demonstrandi quo utitur Jordanus, ut inferius manifestabitur evidenter, et ideo in isto libro magna superfluitas et curiositas continetur". While Boethius and Jordanus apply a *propter quod* explanation, Martianus Capella's treatment of arithmetic in his *De nuptis Philologiae et Mercurii* presents a *quia* explanation, as mathematical notions are therein introduced through a myth or tale (*per modum narrationis*).

51 *Arithm.*, I, 1, 1–7, pp. 6–8, esp. I, 1, 7: "Constat igitur quisquis haec [*scil.* quattuor disciplinae] praetermiserit omnem philosophiae perdidisse doctrinam. Hoc igitur illud quadruvium est quo his viandum sit quibus excellentior animus a nobiscum procreatis sensibus ad intelligentiae certiora perducitur".

52 *Arithm.*, I, 2, 2, p. 11: "Quae cum ita sint cumque omnium status numerorum colligatione fungatur, eum quoque numerum necesse est in propria semper sese habentem aequaliter substantia permanere".

53 *Arithm.*, I, 2, 1, p. 11: "Omnia quaecumque a primaeva rerum natura constructa sunt numerorum videntur ratione formata. Hoc enim fuit principale in animo conditoris exemplar. Hinc enim quattuor elementorum multitudo mutuata est, hinc temporum vices, hinc motus astrorum caelique conversio". On these three Neopythagorean aspects, see Crialesi, "Numbers".

54 John of Holywood, *Algorismus vulgaris*, pp. 1–2: "*Omnia quae a primaeva rerum origine processerunt ratione numerorum formata sunt*, et quemadmodum sunt, sic cognosci habent: unde in universa rerum cognitione, ars numerandi est operativa. Hanc igitur scientiam numerandi compendiosa edidit philosophus nomine Algus unde algorismus nuncupatur, vel ars numerandi, vel introductio in numerum. Numerus quidem dupliciter notificatur, formaliter et materialiter: formaliter ut numerus est multitudo ex unitatibus aggregata; materialiter ut numerus est unitates collectae". My italics. Now, the distinction between Holywood and Boethius becomes evident in their respective conceptions of number, as the former incorporates elements of Aristotelian philosophy in his understanding of number. According to Holywood, number is known both 'formally' and 'materially'. Formally, number is perceived as the multitude arising from the aggregation of individual units. In other words, the form of the number is the coherent assembly of the units that constitute it – *aggregatum* being a technical term denoting the sum. On the other hand, materially, the number is seen as the individual units simply brought together (*collectae*). If one were to rephrase his words, the form of the number is the cohesion of the units that make it up, while the matter of the number consists of the individual units, independent of the aggregation or sum itself. Boethius' Neopythagorean elements, specifically (1), also resurface in other works that can be related to practical arithmetic, such as Thomas Bradwardine's *Treatise on the Ratios*, *Proemium*, 64: "Omne motum successivum alteri in velocitatem proportionari contingit; quapropter philosophia naturalis, quae de motu considerat, proportionem motuum et velocitatum in motibus ignorare non debet. Et quia cognitio illius est necessaria et multum difficilis, nec in aliqua parte philosophia tradita est ad plenum, ideo de proportione velocitatum in motibus fecimus istud opus. Et quia, *testante Boethio, primo Arithmeticae suae: Quisquis scientias mathematicales praetermiserit, constat eum omnem philosophiae perdidisse doctrinam,* – ideo mathematicalia quibus ad propositum indigemus praemisimus, ut sit doctrina facilior et promptior inquirenti". My italics.

55 Martianus Capella, *De nuptiis*, VII, 746, p. 16: "Mihi vero solus numerus approbator qui digitis coercetur; alias quaedam bracchiorum contorta saltatio sit".

56 See *In Calculo*, III, 60–69, pp. 110–115. This will be discussed later in Chapter 6.

57 On the abacus, see Evans, "Schools and Scholars".

58 See Brown, *The Abacus*, 108–124.

59 Ifrah, *Histoire universelle*, especially 239–262 and 344–364; Burnett, "Learning to Write Numerals"; Burnett, "Indian Numerals".

60 Folkerts, "Names and Forms". On the *apices*, see Beaujouan, "Étude paléographique".

61 On fractions in the *Liber abaci*, see Moyon and Spiesser, "L'arithmétique ", especially 405–407. One of the terms Fibonacci uses to refer to fractions is *minutia*, which, as it will be shown in Chapter 6, is precisely Abbo's name standing for fractions. Moyon and Spiesser, "L'arithmétique", 404.

62 Yeldham, “Notation”. There is evidence of the use of fractions also in early medieval glosses to Boethius’ *De institutione musica*. See, for instance, manuscript Paris, BnF, lat. 7202. Evans already pointed out Boethius’ reference to the duodecimal fractions of *semis* $\frac{1}{2}$, *triens* $\frac{1}{3}$, and *bisse* $\frac{2}{3}$ to express the semitone and other tones in continuous proportion. See Evans, “Fractions”.
63 See *In Calculo*, III, 47–54, pp. 101–108. More on Abbo’s treatment of fractions in Chapter 6.
64 The transition from abacus to algorism is explored by Evans, “From Abacus to Algorism”. The differences between these two mathematical traditions are taken from her study.

2 Victorius' *Calculus* and Abbo's *Explanatio*

2.1 Contents of the *Calculus*

The *Calculus* can be divided into three parts: the *Praefatio de ratione calculi*, the series of multiplication tables, and the tables and texts which have been added posthumously.

2.1.1 *Praefatio de ratione calculi*

This text is an introduction to the multiplication tables and it revolves around two thematic cores: on the one hand, the distinction between unit and compounds, and on the other hand, the exposition of the duodecimal parts of the unit as well as the way to use the calculus. As for the first thematic core, the opening section of the *Praefatio* exhibits some philosophical significance. Here, Victorius deals with the status of unity, which lacks the possibility of being divided that is rather proper to all things, as articulated in the following passage:

> The unity from which the entire multitude of numbers proceeds, and which properly belongs to the discipline of arithmetic, indeed admits no division because it is truly simple and subsists with no congregation of parts. As for other things, although there may be something deserving of the name of unity due to its integrity and solidity, yet, because it is composite, it will necessarily be subject to division. For nothing in the entire nature of things, apart from the mentioned unity of numbers, can be found that cannot be wholly distributed by division. This happens because it subsists not on account of simplicity but of composition. For example, we say 'one human being, 'one horse', 'one day', 'one hour', 'one coin', and countless other things of this kind, which, although they have received the name of unity, are nevertheless divided out of necessity of cause and reason. As a summary of this division, the ancients devised such a method of calculation that every integrity to be divided could be rationally cut by partition thanks to it, whether it be a corporeal or incorporeal substance proposed for division.[1]

Victorius explicates the domain to which the calculus is applicable, invoking a distinctive characteristic inherent in all things in the natural realm: composition, and thus, divisibility. Nothing within nature remains exempt from division,

DOI: 10.4324/9781032643472-3

although to human perception everything appears as one and can be defined as 'one thing' due to cohesion and wholeness. This holds true even for 'incorporeal entities', such as temporal units like the 'day' and the 'hour', which fall under the broad category of compounds, too, according to Victorius. Contrarily, what is truly simple and thus stands in opposition to this extensive category of compounds is the unity from which all numbers proceeds – this conceptualisation of unity does not seem to bear any theological strain. As Victorius points out, the unity proper to existing things should be conceived as 'whole', whether it pertains to bodily quantity (what he calls *integritas*) or units of measurement, for both bodily quantity and measures of time can be fractioned, and thus fall under the category of compounds. One could conclude, then, that in Victorius' view, unity does not imply divisions if conceived as the basis for enumerability; whereas if understood as 'whole' (e.g., a day, a human being, etc.), it is subject to partitioning. His calculus, that is, the set of arithmetical tables he developed applies to the domain of unity conceived as a divisible whole, encompassing every entity, be it corporeal or incorporeal, that is not truly simple.

In terms of calculation, a divisible whole is referred to as 'as' or 'axis' and can be fractionated into equal parts (*particulae*). These parts are denoted by specific symbols and derive their nomenclature from their relationship to the integer, that is, the axis. The divisible integer, or axis, is divided on a base 12; consequently, the axis can be mathematically expressed as $\frac{12}{12} = 1$. In the following text, which is part of Victorius' *Praefatio*, for each part of the axis there will be modern fractions in square brackets.[2]

> In this tool [for calculus], the unity is called axis, and its parts, according to their proportion, are designated by their own names. Also, specific symbols have been created for expressing the same words, so that through the differentiation of names and the signs attached to names, the concept of each part is more easily understood. Indeed, the axis, which is represented by the letter I, as one is commonly written in numbers, has twelve parts. If one [part] is subtracted, the remaining eleven parts are called *iabus* ($\frac{11}{12}$). The one that you subtracted, that is the twelfth part, is called an ounce ($\frac{1}{12}$). If two [parts] are taken away, the remaining ten are called *dextans* ($\frac{10}{12} = \frac{5}{6}$), and what you took away, that is two, is called a *sextans* ($\frac{2}{12} = \frac{1}{6}$). If three are removed, the nine that remain are called *dodrans* ($\frac{9}{12} = \frac{3}{4}$), and the three taken away are called a *quadrans* ($\frac{3}{12} = \frac{1}{4}$). If you wish to remove four, the remaining eight are called *bissem* ($\frac{8}{12} = \frac{2}{3}$), and the four taken away are called a *triens* ($\frac{4}{12} = \frac{1}{3}$). With five removed, the remaining seven are called *septunx* ($\frac{7}{12}$), and the five taken away are called *quincunx* ($\frac{5}{12}$). However, when the division is made in the middle, each half consisting of six parts is called *semis* ($\frac{6}{12} = \frac{1}{2}$); an ounce and a half is called *sescuncia* ($\frac{1}{8}$), and half an ounce is called *semuncia* ($\frac{1}{24}$). Now, the remaining small parts [i.e., *minutiae*], by whose combination half an ounce is formed – such as *sicilicus* ($\frac{1}{48}$), *sextula* ($\frac{1}{72}$), and others – will be better understood through the examination of the calculus itself.[3]

2.1.2 *Multiplication Tables*

As Vittorio explains towards the conclusion of the *Praefatio*, the multiplications presented in the tables encompass numbers ranging from 1000 to $\frac{1}{144}$ (*dimidia sextula*). Therefore, these calculations also include the *particulae*, i.e., further duodecimal parts into which the axis and the *minutiae* can be divided.

> The same calculation begins from one thousand and progresses up to fifty thousand, first through multiplication by two, then by three, taking increments through other multiplications, growing with such a multitude of numbers that its sum reaches infinity. It is written with lines descending from the upper part to the lower part, with the upper part containing the sums of the thousands resulting from the multiplication, and the lower part containing the smaller fractions of the divisions [of the axis]. However, in reading [the table], one must start from these [smaller fractions] and go upward until reaching the sum of a thousand, which gradually increases through that multiplication. It should begin with *dimidia sextula* ($\frac{1}{144}$) – through multiplication by two – up to two thousand; then again with *dimidia sextula* ($\frac{1}{144}$) – through multiplication by three – up to three thousand; then with *dimidia sextula* ($\frac{1}{144}$) – through multiplication by four – up to four thousand; and so on until the end.[4]

According to Victorius' own description, his *Calculus* consists of a set of 49 tables, organised into two columns. The left column records the product of multiplication, while the right column contains the corresponding multiplicands. Each table correlates with a multiplier which, although not explicitly indicated, ranges from 2 to 50. The multiplicands as well as their respective products are systematically arranged in descending order (from top to bottom), encompassing a total of 45 numbers spanning from 1000 to $\frac{1}{144}$. Victorius considers only the hundreds (from 900 to 100), the tens (from 90 to 10), and the units (from 9 to 1). Beyond the unit (*assis* or axis), the calculation extends to the multiplication of fractions (*particulae*): from $\frac{11}{12}$ (*iabus*) to $\frac{1}{12}$ (*uncia*). The calculation, then, continues to *minutiae*, namely, the parts of the ounce: $\frac{1}{24}$ (*semuncia*), $\frac{1}{36}$ (*duae sextulae*), $\frac{1}{48}$ (*sicilicus*), $\frac{1}{72}$ (*sextula*), ending with $\frac{1}{144}$ (*dimidia sextula*). To elucidate the structure of the *Calculus*, a representative table is provided, illustrating multiplication by 36.

The multiplications entries are expressed using Roman numerals (from 1000 to 1) and fraction symbols (from $\frac{11}{12}$ to $\frac{1}{144}$). The multiplicands are displayed on the right, in accordance with the previously outlined order. On the left, aligned with each multiplicand, the corresponding products are presented. For instance, considering a table dedicated to the multiplier 36 (as it is the case in Figure 2.1, the initial multiplicand, 1000, is paired with its corresponding product on the left, 36,000. Likewise, the *iabus*, that is, 11 parts of the axis ($\frac{11}{12}$), finds its product on the left as 33.

XXXVI milia	mille
XXXII milia CCCC	DCCCC
XXVIII milia DCCC	DCCC
XXV milia CC	DCC
XXI milia DC	DC
XVIII milia	D
XIIII milia CCCC	CCCC
X milia DCCC	CCC
VII milia CC	CC
III milia DC	C
III milia CCXL	XC
II milia DCCCLXXX	LXXX
II milia DXX	LXX
II milia CLX	LX
mille DCCC	L
mille CCCCXL	XL
mille LXXX	XXX
DCCXX	XX
CCCLX	X
CCCXXIIII	VIIII
CCLXXXVIII	VIII
CCLII	VII
CCXVI	VI
CLXXX	V
CXLIIII	IIII
CVIII	III
LXXII	II
XXXVI	I
XXXIII	ʃʃɣ
XXX	ʃʃʃ
XXVII	ʃɣ
XXIIII	ʃʃ
XXI	ɣ
XVIII	ʃ
XV	ʄɣ
XII	ʄʃ
VIIII	ƴ
VI	ʄ
IIII ʃ	£
III	ɼ
I ʃ	ʅ
I	∪∪
ʃɣ	⊃
ʃ	∪
ƴ	Ψ

Figure 2.1 Table of multiplication on numbers from 1000 to $\frac{1}{144}$ by 36. Re-elaboration of *Calculus*, 27–28.

2.1.3 *Supplementary Tables and Texts*

In the *Praefatio*, Victorius exclusively references the multiplication tables, delineating how his *Calculus* works. Curiously, though, the extant version of the *Calculus* encompasses a broader array of arithmetical tables and diverse short texts on varied subjects. More precisely, this supplementary material includes tables pertaining to addition, subtraction, squares of whole numbers and fractions. Additionally, it features a table illustrating the equivalence of fractions, ounces, and *scripuli*, along with texts addressing the weight of oil and honey (titled *Olearia incipiunt pondera* and *Item mellaria incipient* in Peden's edition). Furthermore, there are tables regarding the units of measurement of geometric surfaces (*De geometria*) and liquid substances (*De rebus liquidis*), as well as concerning the weights of medicinal substances (*Nomina ponderum medicinalium*). This version of the *Calculus* also incorporates an excerpt from Book 16, chapter 25, 1–6 of the *Etymologiae* by Isidore of Seville, presented under the title *De signis ponderum*.[5]

It was probably during the 9th century that this material was added to the *Praefatio* and the multiplication tables (i.e., to the original core of the *Calculus*). A fragment containing parts of the *Calculus* and part of this additional material is preserved in Vatican City, BAV, Vat. lat. 5755, fols. 5r-6v. Notably, in this manuscript, these texts are transcribed by an Irish hand, dating back to the 9th century, establishing the composition and integration of these additional sections into the *Calculus* no later than this period.[6]

2.2 Victorius *scripulorum calculator*: Calculus and the Reckoning of Easter

Calculus (and specifically Victorius' typification) could be regarded as an essential prerequisite for computus itself, namely, the calculation of movable feasts in the Christian calendar. Reckoning skills and liturgical calendar are tightly linked, particularly when considering the historical context of determining the date of Easter within the Christian community. This challenging question captivated Christians, particularly in the initial seven centuries AD, and stemmed from the endeavour to reconcile the Jewish lunar calendar with the Christian solar calendar. The initial predicament emerged from the will to align the celebration of Easter with the Jewish observance of the exodus and liberation from slavery in Egypt, as dictated by the Jewish calendar. This alignment posed theological and computational challenges. While the Jewish celebration could occur on any day of the week, Christian communities, from the 2nd century AD onward, opted to exclusively commemorate the resurrection of Christ on Sunday. Additionally, the Jewish calendar marked the liberation on the 14th day of the first lunation of spring. In contrast, drawing on ancient astronomical principles where spring commenced at the equinox, marked by the sun entering the sign of Aries, Christians adopted this astronomical criterion for determining the Easter celebration.

After the Council of Nicea in 325, the Christian computus started to evolve independently from the Jewish calendar, gradually transforming the Easter computus into an internal challenge within the Christian tradition itself.[7] In the 4th century, a notable divergence emerged in computus models, giving rise to two contrasting approaches. The Western model, embraced by the Church of Rome, and the Eastern model, favoured by the Church of Alexandria, epitomised this dichotomy, hinging on the astronomical precision of the equinox. On the one hand, the Alexandrian system set the equinox on March 21st. This date served as the starting point, that is, the *terminus a quo* for determining the full moon of Easter, with the ensuing calendar limits set at March 22nd and April 25th for the celebration. On the other hand, the Roman system identified March 25th as the equinox, serving as the *terminus a quo* for Easter celebration, and established April 21st as the calendar limit for the observance.[8] The challenge at hand involved devising a luni-solar cycle that could seamlessly incorporate the lunar cycle within the span of a solar year, as well as determining a specific date on the Julian calendar for each lunation and each solar month. In other words, the goal was to create a mathematical system that closely aligned with natural phenomena, establishing a correlation between two standardised astronomical cycles: the solar cycle and the lunar cycle. The standardisation was facilitated by the introduction of the 'full' lunar month of 30 days and the *annus bissextus*. The difficulty laid in the fact that neither the lunations nor the solar year could be accommodated within the total number of days in the Julian calendar as an exact quantity (as an integer number). Moreover, the solar year could not neatly contain an exact (integer) number of lunations. Consequently, two computational methods were conceived to address these discrepancies. For instance, at intervals of every 19 solar years, a number of lunations could be interposed, giving rise to increasingly intricate adjustments to harmonise the two cycles.

In the years 444 and 455, a significant challenge confronted the Church of Rome: the occurrence of Easter beyond April 21st. In an attempt to reconcile the Roman and Alexandrian traditions, Pope Leo I sought assistance from Pascasinus, the bishop of Lilybaeum (modern-day Marsala), and Proterius, the bishop of Alexandria. Despite these efforts, Pope Leo I's hopes for unification were unsuccessful. Consequently, Hilary, who was then the archdeacon of Leo I and later succeeded him as Pope, turned to Victorius of Aquitaine to address and resolve the disparities between the Alexandrian and Roman calendars. In 457, Victorius elaborated the *Cursus paschalis*, a set of computistical tables accompanied by a prologue. These tables projected the date of Easter for the subsequent 532 years, after which the cycle would repeat itself.[9]

Victorius' *Cursus paschalis* fell short of meeting the expectations of Pope Hilary.[10] In spite of this, Victorius' tables gained prominence, especially in the British Isles, after being adopted following the provincial council of Orleans in 541 by churches adhering to the Roman calendar.[11] Starting in the early

6th century, the computus developed by Dionysius Exiguus, a monk originally from Scythia residing in Rome, gradually supplanted Victorius' work. Approximately four centuries later, Abbo himself undertook the reform of the Dionysian computus.[12]

The intricate story about the computus that has been succinctly summarised here bears some insights into Victorius' *Calculus* itself. The reason why pope Hilary turned to Victorius to address the intricate issue of the Easter date may be found in the description of Victorius offered by Gennadius of Marseille. Gennadius, the author of the *De viris illustribus*, a work that completed the series of biographies begun by St. Jerome, likely recognised Victorius' capabilities and expertise in tackling complex calendric challenges, owning to his ability of calculating with fractions – the same fractions that are illustrated and used in his *Calculus*.

> Victorius, of Aquitaine by birth, *calculator of fractions*, was invited by Saint Hilary, the Bishop of Rome, and composed an Easter cycle with the utmost diligence after the four who came before him, namely, Hippolytus, Eusebius, Theophilus, and Prosper. He extended the series of years up to the 532nd year, so that in the 533rd year, the celebration of Easter would resume in the same month and on the same day and with the same moon as the first passion and resurrection of the Lord occurred.[13]

The epithet calculator of fractions (*scripulorum calculator*), attributed to Victorius by Gennadius, underscores his mathematical proficiency, a quality that may have played a crucial role, if not decisive, in his contribution to the Easter computus. The term *scripulus* (which can be expressed as $\frac{1}{288}$ in modern mathematics) originally denoted a unit of weight equal to the 24th part of an ounce ($\frac{1}{12}$), itself constituting the 12th part of an axis.[14] Gennadius employs this epithet due to the arithmetical operations (*calculi*) developed by Victorius in his work, namely, the *Calculus*. As I have already explained, the *Calculus* is a repertory of tables featuring arithmetic operations, including the duodecimal parts of the axis, as well as the *scripuli*. There, Victorius illustrates the methodology for multiplying fractions on a duodecimal basis. Now, proficiency in performing operations on a duodecimal basis was essential for time reckoning within the Julian calendar, where the duration of the day (24 hours) served as the primary unit of measurement.

To calculate time cycles with integer numbers, then, consideration of duodecimal fractions of the day and the corresponding hours was necessary. For example, in Bede's *De temporum ratione* and Abbo's *Explanatio*, terms like *punctus* and *pars* stand respectively for $\frac{1}{4}$ (equivalent to 15 minutes) and $\frac{1}{15}$ (equivalent to 4 minutes) of a solar hour.[15] Bede's *De temporum ratione*, specifically in chapters dedicated to technical aspects preparatory to the study of computus, addresses duodecimal parts (*unciae* or ounces) of the axis, applicable to temporal durations and bodily quantities.[16] To stress the relationship between Bede's and Victorius' expertise, let us know that Charles W. Jones, in the 1943 edition of Bede's computistical works, considered Victorius' *Calculus* as a likely source for Bede's chapter IV (*De ratione unciarum*) in *De temporum ratione*.[17]

Manuscripts such as London, British Library, Egerton 3088 (f. 79r) and Cotton Tiberius E IV (fols. 134v-135r) follow Bede's *De temporum ratione* with two expositions specifically focusing on ounces, further emphasising the significance of duodecimal multiplication for computus.[18] The significance of Victorius' *Calculus* to time reckoning, is also stressed by its manuscript tradition. Many of the surviving copies of the *Calculus* are found alongside other computistical works, often within anthologies dedicated to the ecclesiastical calendar. One notable example is the manuscript compiled by Jacques Sirmond (1559–1651), utilised by Petavius in crafting *De doctrina temporum* (1627). Charles W. Jones identified this manuscript as codex Oxford, Bodleian Library, Bodl. 309. In this specific manuscript, the *Calculus* (fols. 132r-140v) is positioned not after, but between Victorius' *Cursus paschalis* and the tables of Dionysius Exiguus, along with Bede's *De temporum ratione*.[19] This arrangement, thus, is revealing of the interconnectedness of these computistical works. The proximity of the *Calculus* to works by Bede, a major figure in the computistical tradition of the British Isles, suggests a possible link between Victorius' work and the broader context of British computistical practices. This connection becomes particularly relevant in understanding how Victorius' *Calculus* found its way to Fleury abbey, ultimately becoming available to Abbo for further study and commentary.

2.3 Two Hypotheses on the Arrival of the *Calculus* at Fleury

Among the extant copies of Victorius's *Calculus*, it is not possible to identify the specific manuscript that Abbo might have had at hand. Nevertheless, it is possible to put forward two hypotheses on the arrival of the *Calculus* at the library of Saint-Benoît-sur-Loire. The following analysis draws upon three manuscripts predating the 10th century, which transmit the *Calculus* without Abbo's commentary (i.e., the *Explanatio*) but with the supplementary texts and tables:

H: Bern, Burgerbibliothek, 250, fols. 1v-11r (9th c., origin: Fulda or Seligenstadt);[20]

O: Basel, Universitätsbibliothek, O. II. 3, fols. 1v-11r (9th c., origin: Fulda);[21]

P: Vatican City, Biblioteca Apostolica Vaticana, Vat. lat. 5755, fol. 3 (8th-9th c., origin: Ireland and Bobbio).[22]

The first hypothesis (A) builds upon manuscripts H and O, whereas P forms the ground for the second hypothesis (B).

2.3.1 *Hypothesis A: Lupus of Ferrières*

Manuscripts H and O exhibit interconnected traits and can be traced back to Lupus of Ferrières, a monk of the third Carolingian generation distinguished for his intellectual prowess.[23] Lupus of Ferrières, also known as Servatus (805ca.–862ca.), came from Gatinâis, in the diocese of Sens.[24] While the exact date of his entry into the monastery of Ferrières remains unknown, it is documented that from 828 to 836, he studied under Hrabanus Maurus, a disciple of Alcuin of York, at Fulda. Subsequently, Lupus became a 'Palatine cleric' at the court of Ludwig the Pious. Following the emperor's death in 840, he ascended to the position of abbot at Ferrières. Lupus' biography extends beyond monastic life: he served Charles the Bald in military expeditions in Aquitaine, participated as a counsellor in various assemblies, assumed the role of *missus regio* (849), and served as secretary in a number of synods, including those of Soissons (853) and Sens (856). Intellectual pursuits substantially defined his life, as he engaged in debates on predestination, the nature of the soul, and the *visio beatifica* in the 9th century, exchanging with scholars such as Gottschalk of Orbais, Ratramnus of Corbie, Paschasius Radbertus, and Hincmar of Reims.

Of particular relevance to our inquiry is Lupus' time at Fulda and later as abbot at Ferrières. Following his studies with Hrabanus Maurus, Lupus gained renown for his scholarly and teaching abilities. It is plausible that students from other religious and learning centres, like Heiric of Auxerre (841–876), sought instruction at Ferrières under his guidance. Evidence of a *schola* with a structured didactic programme at Ferrières at Lupus' time can be inferred from his correspondence, notably *Epistola* 115*bis*, in which Lupus laments the death of an advanced-level student.[25] Ferrières emerged as a significant hub within a network of institutions, including Fleury, as evidenced by some annotations found in a manuscript containing Lupus' epistles, namely, Paris, BnF, lat. 2858. These annotations are made by a 'Fleurisian hand' and in all likelihood belong to an anonymous student of Lupus who also glossed other classical texts originating from Fleury and Auxerre.[26] The same hand from Fleury, for instance, appears also in the manuscript Paris, BnF, lat. 5763, transmitting the *De bello gallico* used by Lupus for his teaching at Ferrières.[27] A certain link between Ferrières and Fleury, and between Lupus and Fleurisian students thus, is evidenced in some extant manuscripts.

Among Lupus' 'polygraphical letters' (that is, those letters encompassing diverse topics), one notable correspondence dates back to 836. This letter, written from Fulda, is addressed to Einhard (775–840), abbot of Seligenstadt and author of the *Vita Karoli Magni*. In this communication, Lupus apprises Einhard of his delayed departure for Seligenstadt and anticipates forthcoming discussions on some issues upon his arrival. These issues include passages from Boethius' *De arithmetica* which Lupus finds difficult to understand (i.e., I, 4; I, 31; II, 25), along with grammatical intricacies concerning the accentuation of particular words.[28] Moreover, Lupus expresses his eagerness, under Einhard's guidance, to embark on the study of Victorius of Aquitaine's *Calculus*: "In Victorii quoque calculum praevia Dei gratia vestraque doctrina ingredi cupio" ("I also wish, with the prior grace of God and your teaching, to delve into the *Calculus* of Victorius of Aquitaine").[29] Lastly, Lupus apologises to Einhard for the inability to promptly return the codex containing Aulus Gellius' *Noctes atticae*, as Hrabanus Maurus, the abbot of Fulda, has retained it for copying. Nevertheless, Lupus assures Einhard that during his upcoming journey to Seligenstadt, he will personally return the codex along with any others he has borrowed.[30]

From these concise details, three significant pieces of information emerge regarding Victorius' *Calculus* and its circulation. (1) In the first half of the 9th century, the *Calculus* was present in the scriptorium of Seligenstadt – manuscript H, which contains the 'supplementary material' of the *Calculus*, suggests that the *terminus ante quem* for their composition could be advanced to the year 836, corresponding to the date of Lupus' letter. (2) Einhard exhibited a mastery of the *Calculus* and Boethius' *De arithmetica*, as evidenced by Lupus' request for further explanations. (3) The exchange of manuscripts between the two Germanic monasteries, Seligenstadt and Fulda, was a longstanding tradition. This practice facilitated the sharing of intellectual resources, contributing to the dissemination and accessibility of scholarly works such as the *Calculus* within these monastic centres.

As previously mentioned, the letter Lupus addresses to Einhard is dated to 836, representing the concluding period of Lupus' stay at Fulda. Given that Lupus developed a keen interest in arithmetic and calculation shortly before departing Fulda, it is conceivable that he brought a copy of Victorius' work with him to Ferrières. Consequently, the *Calculus* may have become known to Lupus' students, including the anonymous Fleurisian who annotated Lupus' teachings. The introduction of the *Calculus* at Fleury via Ferrières thus appears plausible, also considering that manuscript H, today preserved in Bern, eventually found its way into the collection of Pierre Danielle of Orléans (1530–1604) and then of Jacques Bongars (1554–1612), which housed many manuscripts from Fleury.[31]

2.3.2 *Hypothesis B: Columbanus and the Abbey of Bobbio*

Between the 7th and the 8th centuries, the Christian communities of the British and Irish islands were divided on the issue of the dating of Easter. The division arose between those adhering to the Roman Church tradition and those aligning with the principles established by Anatolius of Laodicea (d. 282), which disseminated in these regions through the *Latercus* (an Easter table with an 84-year cycle).[32] Relying on the *Vita Wilfridi* (c. 720) by Stephen of Ripon, a 'Roman faction' emerged in this dispute, advocating for the tables developed by Victorius of Aquitaine.[33] Further supporting evidence of the dissemination of Victorius' Paschal tables in this area is found in the codex Bremen, Staats-und Universitätsbibliothek, 46, originating from the monastery of St. Gallen and dating to the 10th century.[34] It serves as an anthology of computistical texts studied in Ireland between the 7th and the 8th centuries, transmitting a prologue of Irish origin (dating to 669) to Victorius' *Cursus Paschalis*.[35] As previously explained, the multiplication tables presented in Victorius' *Calculus* could be considered preparatory to the challenges related to the ecclesiastical calendar. Therefore, it is plausible that the *Calculus* was studied in this insular context alongside the *Cursus Paschalis*.

To reconstruct a conjectural trajectory of the *Calculus* from Ireland to France, it is noteworthy that Columbanus (543–615), missionary and advocate of the Irish monasticism, was familiar with Victorius' computistical work. In a letter addressed to the Frankish bishops of the synod of Chalon, Columbanus explicitly contrasts the more recent computistical tradition of Victorius of Aquitaine to the earlier Celtic one, rooted in Anatolius of Laodicea.[36] Now, it is essential to recall that in manuscript P, the oldest extant witness of the *Calculus*, the *Calculus* is copied by an Irish hand and is preceded by an anonymous text on Easter computation. P only preserves the final part of the *Calculus* (i.e., the multiplication tables from ×41 to ×50), and the 'supplementary tables' of addition, subtraction and signs of the duodecimal parts of the axis. According to Elias Avery Löwe, P was copied in Ireland between the 8th and 9th centuries and was later transported to Bobbio in Northern Italy, where some fragments served as reinforcement for ligatures.[37] Today, P appears as a palimpsest, with an 11th-century copy of Augustine's *De Trinitate* written over the original text. The path of the *Calculus* in P – from Ireland to Italy – closely mirrors the path of Columbanus, who left the monastery of Bangor in 590 to embark on European travels.[38] Columbanus resided in France until 610, founding the Luxeuil monastery, and then journeyed through Switzerland, where his disciple Gallus established a monastery bearing his own name. In 613, Columbanus halted in Milan, at the court of the Lombard king Aigulf, who granted him the territory of Bobbio in 614 to establish an ecclesiastical community. Giovanni Mercati dismissed the idea that Columbanus brought the codices forming the original Bobbio library from Ireland.[39] However, the relationships Columbanus established between the continent and the islands, along with the foundations of significant monastic centres in France and Switzerland by him and his disciples, do not allow us to rule out that the *Calculus* may have traversed these channels, ultimately reaching Fleury – and in this respect, Pierre Riché has already shown the established exchanges between the monastery of Fleury and the Celtic countries.[40] To bolster his arguments, the foundation record of the Fleury-sur-Loire monastery in 651, preserved in abbot Leotebold's will, specifies that the monks would adhere to the rules of St. Benedict and St. Columbanus.[41]

The manuscripts that served as a trail for the formulation of the two hypotheses do not provide conclusive evidence for identifying the exemplar of the *Calculus* used by Abbo. Assuming that the *Calculus* consulted by Abbo has survived and remained undiscovered, there is a potential clue, noted by Alison Peden, that could aid in its identification.[42] This clue lies in Abbo's own commentary, the *Explanatio*, particularly in the way he interprets Victorius' expression "et nulla partium compositione subsistit".[43] Regarding the word *compositio*, Abbo warns his students that in certain codices ("in quibusdam codicibus") *congregatio* could be read instead of *compositio*. This disclosure indicates two important points: first, that Abbo consulted different versions of the *Calculus*, and second, in his own copy, the correct expression "et nulla partium compositione subsistit" was present. Unfortunately, his specific wording is not found in the manuscripts that exclusively contain the *Calculus*. Manuscripts O and P lack the *Praefatio*, while H presents *congregatio* instead of *compositio*.

2.4 Abbo's *Explanatio*: Manuscripts and Critical Editions

With a (hopefully) clearer understanding of the *Calculus*' contents and 'material history', I now shift the focus to Abbo's commentary, specifically examining relevant aspects of the manuscript and printed tradition. Alison Peden's critical edition of Abbo's *Explanatio*, published in 2003, presents a bipartite *stemma codicum*. The manuscript Berlin, Staatsbibliothek, Phill. 1833 (Rose 33) – hereafter referred to as F – and the Vatican, BAV, Reg. lat. 1281 – hereafter A – constitute a branch of the stemma, closely related to the archetype. They date to the late 10th and early 11th centuries, respectively, both being originally copied at Fleury. Their shared origin is evidenced by details such as the use of the spelling of *loyca* instead of *logica*, which is typical of the Fleurisian scriptorium. F is an anthology of 'Abbonian texts' on computus. Valentin Rose suggested that this codex was assembled under Abbo's supervision, yet Peden cautions against this view, for she notes that the *Calculus* transmitted with the *Explanatio* in F is incomplete, making it unlikely that Abbo did not have a complete version at hand.[44] Manuscript A, too, compiles 'Abbonian material' on computus and logic.[45]

The other six manuscripts considered by Peden constitute the opposite branch of the stemma, all descending from a common source dated later than A and F. Among these are Karlsruhe, Badische Landesbibliothek, 504 (hereafter K), Bruxelles, Bibliothèque Royale, 10078 (hereafter G), and Vienna, ÖNB, 2269 (hereafter V). Examining the contents of manuscripts K and G provides insights into the intellectual context in which Abbo's *Explanatio* circulated, indicating its potential relevance in the field of musical theory. In K, dating to the 12th century from the monastery of St. Michael of Bamberg, the *Explanatio* is preceded by Berno of Reichenau's *Prologus in tonarium* and Guido of Arezzo's *Micrologus*.[46] Similarly, in G, the *Explanatio* is embedded within a collection of texts on music theory, including the second part of the *Disputatio in Somnio Scipionis* by Favonius Eulogius, the *Scholica enchiriadis de musica* and other texts on musical modes and consonances. Van de Vyver believed this manuscript to belong to the chronicler Sigebert of Gembloux (1030–1112), who in his *De scriptoribus ecclesiasticis*, considered the *Explanatio* as a manifestation of Abbo's mastery of all liberal arts.[47] Notably, manuscript V, from the 11th century, is a comprehensive anthology on the liberal arts. It transmits works by Boethius – such as his logical works, along with *De arithmetica* and *De institutione musica* –, an anonymous treatise *De geometria*, Hyginus's *Astronomia*, but also philosophical works such as Calcidius' *Commentary on the Timaeus* and Macrobius' *Commentary on the Dream of Scipio*.[48] The arithmetical section of this anthology comprises Boethius' works and Abbo's *Explanatio*. Following these, a short text (fols. 140ra-140va) titled *Regulae eiusdem de minutiis*, likely to be attributed to Abbo himself, is included.[49]

As one might expect, the history of the critical editions of the *Explanatio* is closely intertwined with that of Victorius' *Calculus*. Initially, the *Calculus* was included in the Patrologia Latina among Bede's spurious works with the title *De ratione calculi libellus*, containing only the *Praefatio* and part of the multiplication tables.[50] In 1863, it was attributed to Victorius of Aquitaine and published as *Victorii argumentum calculandi* by Wilhelm Christ.[51] In 1871, it was edited and published, still incomplete, by Gottfried Friedlein.[52] Today, a complete and more intelligible version is available in Peden's edition.

The prologue of the *Explanatio* (i.e., chapter I of Peden's edition) was first published in 1717 by Edmond Martène and Ursin Durand in the first volume of the *Thesaurum novum anecdotorum*.[53] Later, it was included in the Patrologia Latina with the title *Praefatio Commentarii in Cyclum Victorii*.[54] In 1863, Wilhelm Christ added parts of the final section of the *Explanatio*, nevertheless arranging them in a non-sequential way, for his intention was only to provide an understanding of the final part of Victorius' *Calculus* precisely through Abbo's commentary. Reflecting this purpose, the title of the edition of Abbo's text is *Victorii Calculi pars altera ex commentariis Abbonis Floriacensis adumbrata*.[55]

Moreover, in Nikolai Bubnov's edition of Gerbert of Aurillac's mathematical works – more specifically in the section dedicated to the *Abacistae Gerberto aequales* – one can find a transcription, mainly based on manuscript F, of additional passages from the *Explanatio*, in which Abbo dwells on Victorius' multiplication tables.[56] In Bubnov's view, these passages reveal Abbo's limited skills as an *abacista*, that is, as an arithmetician. Bubnov references verses found at the end of the *Explanatio* in manuscript G: "Nunc manibus fer aquas quia sat iam sumpsimus escas/Qui serit in lacrimis, recipit hic gaudia messis/Hic abbas abaci doctor dat se Abbo quieti".[57] Despite Abbo's expertise proclaimed by the scribe, Bubnov concludes negatively: "Although Abbo boasts of being a master of the abacus, he seems to have been not sufficiently skilled in this art" ("Quamquam Abbo se abaci doctorem esse gloriatur, huius artis tamen non satis peritus extitisse videtur").[58] Additionally, Bubnov included under the title *Abbonis abacus* a multiplication table found in manuscript Oxford, St. John's College, 17. This table is similar to the ones we find in the *Explanatio* and in

Byhrtferth's *Enchridion*, but is supplemented with rules for multiplication using fingers.[59]

Editions of Abbo's *Explanatio* previous to Peden's provide only a partial and fragmentary understanding of the text, focusing primarily on its reckoning aspects. These editions did not fully capture the general sense, philosophical depth, or strictly mathematical aspects of Abbo's commentary. These aspects along with Abbo's mastery of the quadrivial arts can be finally savoured in Peden's edition.

Peden's edition divides Abbo's commentary into three chapters of varying lengths. The first chapter includes the 'first prologue' (section I, 1–3 in the edition); the second corresponds to the 'second prologue' (II, 1–16); and the third encompasses the remaining part of the *Explanatio* (III, 1–100), covering the commentary on Victorius' *Praefatio* and tables. Peden further divides each chapter into paragraphs. However, this division does not seem to be driven by specific philological reasons, since it is not consistent with the manuscript tradition. For all manuscripts transmitting the *Explanatio* share a distinct partition marked by decorated initials – which do not correspond to the chaptering proposed by Peden. For example, in codex F, the text bears four ornate initials plus a simple one, which are distributed in the following way:

i On fol. 7v, a zoomorphic initial forming the 'C' (for *Calculus*) is decorated with floral motives, containing the drawing of an animal, likely a goat (i.e., the start of *In Calculo*, I, 1).
ii In the right column of the same fol. 7v, an anthropomorphic initial forming the letter 'A' (for *Amor*) is decorated by geometric and floral motifs, featuring a human head with an open mouth, presumably in the act of speaking (i.e., the start of *In Calculo*, II, 1).
iii On fol. 9r, another zoomorphic initial forming the 'U' (for *Unitas*) is decorated with floral motifs, containing a drawing of a fox (i.e., the end of *In Calculo*, II, 16).
iv On fol. 16v, a monstrous animal with a bird body and two dog heads forms the letter 'E' (for *Expeditis*) (i.e., the start of *In Calculo*, III, 59).
v On fol. 20r, there is a simple penwork initial for the letter 'D' (for the particle *De*) (i.e., the start of *In Calculo*, III, 91).

The arrangement of these ornate letters suggests a partition of the text that is more consistent with the contents and logical-argumentative structure of the work. In the following discussions focusing on the contents of the *Explanatio*, I will stick to this division into five sections.

2.5 Contents of the *Explanatio*

The chaptering that I have just uncovered based on the manuscript tradition, helps us to identify five sections of the *Explanatio*. The first consists of a prologue to the commentary (i.e., chapter I in Peden's critical edition).

The second stands as a theological premise to Abbo's interpretive endeavour (i.e., II, 1–16). The third (i.e., III, 1-III, 59) is the actual commentary on the prefatory text of Victorius' *Calculus*. The fourth (i.e., III, 60-III, 91) gathers Abbo's commentary on Victorius' multiplication tables as well as on the supplementary ones. The fifth (i.e., III, 92-III, 100), which one might define as the 'physical' section, presents a discussion of the specific weights of two liquid substances, oil and honey, still based on the supplementary tables in the *Calculus*.

2.5.1 The Prologue (Section I)

Abbo provides an *accessus*, a brief introduction to Victorius' work, where he outlines the motivations behind composing a commentary on the *Calculus*. There are two main reasons: firstly, the requests of his fellow monks in Fleury seeking for a clarification of the contents of the *Calculus*; and secondly, his personal dedication to the liberal disciplines, especially arithmetic.

2.5.2 The Theological Premise (Section II)

Abbo connects the study of calculus to a specific branch of philosophy, the *quadrivium*, which encompasses four mathematical disciplines: arithmetic, geometry, music, and astronomy. The calculus appears to extend to reality as a whole, insofar as everything is created by God in accordance with number, measure, and weight, as stated in the biblical verse *Wisdom* 11:21.

This section of the text, referred to as 'theological premise', aims to elucidate the nature and causal significance of number, measure, and weight mentioned in the biblical verse, mainly on the basis of Claudianus Mamertus' *De statu animae*.[60] The content discussed in this section might give the impression of an independent text, especially considering that Abbo introduces it with another title, namely, *Tractatus de numero, mensura et pondere*, which is also referenced by Abbo in his *Quaestiones grammaticales*.[61]

2.5.3 The Commentary on Victorius's Praefatio *(Section III)*

The structure of this section is shaped by the *lemmata* of Victorius' *Praefatio*. While commenting on it, Abbo allows for extensive digressions prompted by the occurrence of specific *lemmata* or topics. These interventions highlight the originality of Abbo in relation to the author he is commenting on – a task that he views as precisely the duty of a commentator.[62] Tools and notion taken from different disciplinary fields are used by Abbo to elucidate Victorius's prefatory text. The intertwining of the liberal arts is especially aimed at providing Abbo's brethren with an introduction to arithmetic. This aligns with a pedagogical practice used by early medieval masters, especially in introducing their students to Boethius' *De arithmetica*.[63] Abbo's intent is also evident in the alternative title he himself gives to his *Explanatio*, namely, *Ysagoge arithmeticae*. The *Explanatio*, then, is really an introduction to the study of both speculative and practical arithmetic, that is, number theory and calculus.

Within this isagogic framework, logic and physics represent the other main areas of inquiry on which Abbo relies for his commentary.[64] Logic serves as a tool supporting the analysis and understanding of the discussion about arithmetic, while physics (understood as knowledge of supra- and sub-lunar phenomena) expands the boundaries of such discussion by relating the specific contents of arithmetic to the realm of natural reality. Accordingly, astronomy, too, plays a significant role in the commentary, also helping to understand how incorporeal entities, such as the day and the hour, can be divided. Abbo does not hesitate to resort to the disciplines of rhetoric and grammar to occasionally clarify both the structure of the *Preface* and some of Victorius' expressions therein. Moreover, the dense network of references to Latin classical literature – frequent are the quotations from Horace, Juvenal, Virgil, Sallust, and Terence – showcases Abbo's erudition. Thus, the thoughts gathered in this section build a map of Abbo's liberal knowledge, which is not limited to arithmetic.

The philosophical ideas on which Abbo expounds more extensively revolve around the concepts of unity – absolutely simple when referred to God but divisible when understood as a whole ascribed to every created entity – and composition – an intrinsic characteristic of everything subject to change. Around this philosophical binomial, Abbo articulates and develops his unique philosophy.

2.5.4 *The Commentary on Victorius's Multiplication Tables (Section IV)*

This section involves Abbo's explanation of Victorius' multiplication tables and the supplementary material in the *Calculus*. Abbo does not restrict himself to a straightforward explanation of how these arithmetical tables function. Instead, he offers an interpretation that draws on Boethius' number theory, matching speculative with practical arithmetic. An extensive exposition of the inequality ratios which is specifically borrowed from the *De arithmetica* is provided in this section.[65]

2.5.5 *On Qualitative Physics (Section V)*

Abbo comments only on the spurious text of the *Calculus* that deals with the equivalences between the weight measures of oil and honey.[66] If in the preceding sections, Abbo aimed to establish an introduction to arithmetic, in this final section of the commentary, he engages in a discussion on sublunar cosmology and 'qualitative physics' of the elements. This discussion appears to be linked to a Peripatetic tradition, as Abbo relies on the pseudo-Aristotelian *Problemata* transmitted through Macrobius' *Saturnalia*.[67]

2.6 The Pedagogical Aim and the Role of the Commentator

Before delving into the philosophical contents, arithmetical theories, and physical discussions presented in the *Explanatio*, let us consider some preliminary aspects by addressing the following questions: For whom was this work intended? What was its aim? And in what way is the author supposed to fulfil this goal? Abbo answers these questions in the initial part of his commentary. His intention is to assist those not skilled in reckoning to master this kind of knowledge, by 'building a bridge' to arithmetic, that is, by providing an introduction to this discipline in the form of a commentary.[68]

Besides specifying his aim and the intended audience for the text, Abbo also details the manner in which he will comment on Victorius' *Calculus*: he will not leave unexplained what Victorius openly expounded as well as what he has not covered.[69] Abbo's approach is not merely to provide a description of the contents of the *Calculus* elucidating practical arithmetic, but to initiate his brethren into speculative arithmetic, too. This involves delving into the theory of numbers as articulated by Boethius in his *De arithmetica*, while also clarifying theoretical assumptions that Victorius did not explicate.

Such is the way Abbo proceeds in glossing the *Calculus*. In this respect, it is important to note that Abbo's analytical explanations only partially fit the definition of 'glosa' offered by Isidore of Seville in his *Etymologies* and align more closely with the one provided by William of Conches in the 12th century. According to Isidore, the glossator's work is simply to clarify the obscure meaning of words by replacing it with a clearer one.[70] In contrast, William describes the gloss as intended to grasp the meaning of a text ("solam sententiam exequens") as well as to elaborate on it and somehow go beyond the doctrine of the author commented on ("continuatio vel expositio litterae").[71] Abbo's adherence to William's paradigm is not surprising, considering that this approach is to be found also in glosses composed in the central centuries of the Middle Ages, especially the 12th one.[72] What is striking, though, is Abbo's choice of the text to be glossed: instead of selecting a strictly philosophical or literary text – in the 10th century we would expect something like Martianus Capella's *De nuptiis* –, Abbo opts for a short technical text on calculus, which serves as a bridge to the theoretical and practical foundations of arithmetic.

Abbo's *Explanatio* displays another element that ties it to the literary genre of gloss developed from the 9th century onward: it is the *accessus ad actorem* appearing in the first section of his text, where Abbo makes a digression on Victorius and introduce the *Calculus*. This practice can be traced back to late antique Greek commentators of Aristotle's logical works and to the Ciceronian rhetorical tradition, both of which influenced medieval literary culture through Boethius.[73] In the first version of the *Commentary on Porphyry's Isagoge*, Boethius answered six preliminary questions about the work and its author: *intentio*, *utilitas*, *ordo*, *an sit germanus eius cuius opus esse dicitur, inscriptio, ad quam partem philosophiae.*[74] In the second version of the same commentary, as well as in the one on Aristotle's *Peri hermeneias*, he focused on two: *intentio* and *utilitas.*[75] Differently, in his *Commentary on Aristotle's Categories* he dwelled on the following aspects: *intentio, utilitas, ordo* e *pars philosophiae.*[76] Evidently enough, the number of questions constituting an *accessus* is not fixed. This is so for other early medieval authors who happened to adopt different patterns for the *accessus* in their commentaries. For instance, in John Scotus Eriugena's *Annotationes in Marcianum*, the following elements are presented: a brief introduction to the author Martianus Capella, the subject-matter of the work, and a concise explanation of the personifications of the two main characters, Philology and Mercury.[77] Commenting on the same work, Remigius of Auxerre focuses on the so-called *circumstantiae*: *quis, quid, cur, quomodo, ubi, quando, unde.*[78] In his own *accessus*, Abbo addresses only Victorius' intention and the usefulness of his *Calculus*. Abbo's choice, thus, aligns with Boethius' emphasis in his commentaries on Porphyry's *Isagoge* and Aristotle's *Peri hermeneias*. Let us see how Abbo explores these two preliminary aspects.

> At that time, Victorius' intention was that the reader could multiply and divide the sums of numbers with accuracy; or that something be proposed concerning the number-related arts, such as arithmetic, geometry, music and astronomy; or that measure and weight, which are of interest to those versed in calculus, be discussed. [...] However, since all things are created in number, measure, and weight, it is appropriate to consider each of these things individually so that, having observed their nature, each of the things in which they are involved as causes can be more easily understood.[79]

According to Abbo, Victorius' intention in composing the *Calculus* was to impart proficiency in arithmetical operations and measurement practices – and collaterally, in those disciplines grounded in the nature of number. Yet, Abbo's view on the significance of Victorius' work becomes more apparent when considering the biblical passage *Wisdom* 11:21, stating that everything was created according to number, measure, and weight. Abbo contends that the need to delve into and clarify the nature of these three elements – collaterally to Victorius' discourse – is that they can be recognised as exemplary causes of all things. In essence, Victorius' text not only serves a technical purpose (i.e., to master calculus) but also presents theological and metaphysical implications, a topic that will be explored further in the subsequent chapter.

There is another element that Abbo highlights in his *accessus* and it consists in the Ciceronian rhetorical strategy employed by Victorius in his *Praefatio*. In Abbo's eyes, Victorius effectively captured the reader's attention, fostering a sense of benevolence, attentiveness, and receptiveness. In Abbo's perspective, this was achieved by positioning himself as someone who drew upon the wisdom of the ancients, by previewing the expansive scope of his work, and by adopting a style of discussion that was concise and straightforward.

> As is customary in a preface, with remarkable brevity, [Victorius] captured the benevolence, attention, and docility of the listener: benevolence, by mentioning that, out of humility, he does not portray himself as the inventor of the subject of calculus but as a follower of the ancients; attention, by declaring that he is about to discuss the division of things into incorporeal and corporeal; docility, by promising to handle a concise division with clarity and openness and avoiding weariness (being about to discuss such grand matters of length).[80]

As just mentioned, Abbo looks at Victorius' approach through the lens of Ciceronian rhetoric. In *De inventione* and *Topica*, Cicero distinguished different parts of an oratorical speech. The first part of the speech (i.e., the *exordium*) was divided into *principium* and *insinuatio*. The former aimed to create precisely goodwill in the audience. Cicero outlined rules for this, including avoiding arousing envy, making the topic interesting, and maintaining clarity and brevity in the speech. By sticking to these principles, the speaker could cultivate a benevolent, attentive, and docile audience.[81] This Ciceronian rhetorical structure was transmitted to the medieval Latin tradition with little deviation. Both Boethius and Martianus Capella, for example, presented the same three kinds of dispositions that a speaker could produce in the hearer.[82] Thus, Abbo's explanation of the three rhetorical elements – benevolence, attention, and docility – clearly has a Ciceronian origin. This rhetorical foundation is also supported by manuscript evidence. There is a significant collection of Boethian and Ciceronian writings on logic and rhetoric that was likely assembled under Abbo's supervision, found in manuscripts such as Leiden, Bibliotheek der Rijksuniv, Vossius lat. F 70, I; Orléans, Bibliothèque municipale, 277; and Paris, BnF, nouv. acq. lat. 1630.[83] Along with works by Apuleius, Pseudo-Augustine and Macrobius, this collection includes Cicero's *De Inventione* and *Topica*, together with Boethian commentaries on these works. Having explored the Ciceronian influence on the writing of the *accessus* in the *Explanatio*, let us now briefly turn to Abbo's conception of the commentator's role, as detailed in the text below.

> Ancients commented on – because it pleased the philosophers to reduce everything infinite and doubtful at least to the level of likelihood through a form of demonstration, so that they could derive some speculation. For they could not comprehend any knowledge of the infinite, which is null. However, when he [Victorius] says "they commented on", that is, they made it up (*finxerunt*), it seems to suggest that they did not grasp the true nature of reality. This is because 'make up' (*fingere*)

usually implies not only composing but also putting forward a kind of pretence. In this context, however, 'they commented on' signifies 'invented' or 'devised' – as in the words of Terence: 'manage, fabricate, invent' [*Andria*, II, 334] – even though fiction, that is, pretence, is an imagining of the truth. Hence, we call them commentators, those who, by inventing many plausible explanations, illuminate a truth somehow veiled in obscure sentences (*involuta veritas*), and these invented explanations are called 'commentaries'.[84]

The role of commentators, according to Abbo, involves unravelling the authentic meanings hidden behind the obscure expressions of philosophers. Thus, commentators elucidate the philosophers' obscure sentences by introducing analogies, or even fictions. Once again, the concept of the commentator in Abbo's work aligns with the perspective expressed by William of Conches and shared by some 12th-century authors.[85] The expression *involuta veritas* in Abbo's text echoes the notion of *integumentum* used by 12th-century Chartrian masters, suggesting that profound truths can be concealed behind the literary genre of myth, acting as a kind of 'wrapping' (*involucrum*).[86]

The roots of Abbo's conception of the commentator can be traced to two potential sources. Firstly, it draws from Cicero's rhetoric, specifically his work *De Oratore*. In the exchange between two characters of the dialogue, Cotta and Crassus, Cicero introduces the binomial *involucra atque integumenta*. To Crassus' view on eloquence, Cotta replies that he has retained nothing but traces (*vestigia*) of Crassus' discourse, explaining what he did retain by means of an analogy: he feels like someone who enters a luxurious house and yet does not see fabrics, nor silverware, nor furniture as they have been hidden. In the same way, Cotta grasped that Crassus' speech was intellectually valuable, but only as though through a veil – the veil, thus, having a negative value for Cotta.[87]

Macrobius' *Commentary on the Dream of Scipio* is another possible source, for there, it is explained that a specific type of fiction (*narratio fabulosa*) serves not solely for pleasure but for conveying complex ideas, and can therefore be employed in philosophy when the subject transcends human linguistic capacity, necessitating the use of analogies.[88] It is especially the alignment with Macrobius that emerges in an 'Abbonian florilegium' transmitting the *Commentary on the Dream of Scipio*. When it comes to the passage regarding the "narratio fabulosa", the excerpt from the florilegium recites: "De Fabulis. Nec omnibus fabulis philosophia repugnat ... attingunt, sed ad similitudines et exempla confungiunt, quia summus deus nataque ex eo mens sicut ultra animam uta (sic) supra naturam sunt".[89] Such an excerpt emphasises the connection between Abbo's perspective and Macrobius' explanation of the use of fiction in philosophy: philosophy does not reject all fables, and they are not incompatible. Instead, philosophers (and commentators) touch upon them and use them to connect with similarities and examples, in order to make sense of issues that transcends nature, such as God and God's mind.

Both the Ciceronian and Macrobian explanations serve as a background for Abbo's idea of the commentator's role and the type of reading he intends to offer to his audience. This background becomes evident as Abbo unveils theological issues omitted by Victorius. To such 'theological premise' the next chapter is devoted.

Notes

1 Victorius, *Praefatio*, in *In Calculo*, 3 (hereafter I report only the pages of Peden's critical edition): "Unitas illa, unde omnis numerorum multitudo procedit, quae proprie ad arithmeticam disciplina pertinet, quia vere simplex est et nulla partium congregatione subsistit, nullam utique recipit sectionem. De ceteris vero rebus, licet aliquid tale sit, ut propter integritatem ac soliditatem suam unitatis meruit vocabulo nuncupari, tamen quia compositum est, divisioni necessario subiacebit. Nihil enim in tota rerum natura, praeter memoratam numerorum unitatem, tam unum inveniri potest, quod nulla omnino valeat divisione distribui. Quod ideo fit, quia non simplicitate sed compositione subsistit. Dicitur enim unus homo, unus equus, unus dies, una hora, unus nummus et alia huiusmodi innumerabilia, quae licet unitatis sint sortita vocabulum, tamen pro causae atque rationis necessitate dividuntur. Ad huius divisionis compendium tale calculandi argumentum antiqui commenti sunt, ut omnis dividenda integritas rationabili per illud possit partitione secari, sive id corpus sive res incorporea sit quod dividendum proponitur".

2 For the duodecimal system of fractions, see Chapter 6.

3 *Calculus*, 3–4: "In hoc argumento unitas assis vocatur, cuius partes iuxta proportionalitatem suam propriis sunt insignitae vocabulis. Notis etiam ad hoc excogitatis, per quas eadem vocabula exprimantur, ut per discretionem nominum et notas nominibus affixas uniuscuiusque particulae notio facilius advertatur. Et assis quidem, qui per litteram I, sicut in numeris unum scribi solet, exprimitur, XII partes habet. Quarum, si unam ei detraxeris, reliquae undecim partes iabus dicuntur. Illa vero quam detraxisti, id est duodecima, uncia vocatur. Si duas sustuleris, decem residuae dextans et quod sustulisti, id est duae, sextans appellatur. At si tres dempseris, novem quae remanserunt dodrans et tres demptae quadrans vocantur. Quod si quattuor tollere velis, octo reliquas bissem et quattuor trientem nominabis. Quinque vero sublatis, septem residuas septuncem et quinque sublatas quincuncem placuit appellari. Cum vero per medium fuerit facta divisio, utrumque dimidium senis partibus constans, semissem vocitarunt; unciam autem et dimidiam, sescunciam, unciaeque dimidium semunciam. Iam reliquae minutiae, quarum congestione dimidium unciae conficitur, ut sunt sicilici, sextulae et cetera, melius ex ipsius calculi inspectione cognoscuntur".

4 *Calculus*, 4: "Incipit autem idem calculus a mille et usque ad quinquaginta milia progreditur, primo per duplicationem, deinde per triplicationem, tum per ceteras multiplicationes incrementa capiens tanta numerositate concrescit, ut usque ad infinitum quantitatis eius summa perveniat. Scribitur vero lineis a superiori parte in inferiorem descendentibus superius milium summas ex multiplicatione venientes, inferius divisionum minutias continentibus. A quibus tamen in legendo principium est faciendum et sic sursum versus eundum, quousque ad miliarii summam, quae ex illa multiplicatione paulatim adcrescit, legendo venitur. Incipiendumque a dimidia sextula per duplicationem usque ad II milia, inde iterum a dimidia sextula per triplicationem usque ad III milia, tum a dimidia sextula per quadruplicationem usque ad IIII milia et sic usque ad finem".

5 *Calculus, Additional Texts* 1–11, pp. 38–62. As explained in the final chapter of this book, Abbo directs his primary focus towards the texts related to the weight measurement of oil and honey, exploring them through the lens of what is referred to as 'qualitative physics'.
6 Löwe, *Codices latini antiquiores, Supplement,* 1734, 19; Bischoff, *Mittelalterliche Studien*, 40. The link between the *Calculus* and the Irish area will be clarified in the following sections of this chapter.
7 Cf. McCluskey, *Astronomies*, 77–84. For a survey of the constitution of the Christian calendar from its origin to the Venerable Bede, see Wallis, "Introduction", xxxiv–lxiii.
8 Van De Vyver, *L'évolution du comput.*
9 The *Cursus paschalis* and the correspondence between Victorius of Aquitaine and Hilary are edited in Kursch, *Studien*, 16–52. In the *Cursus*, the dating of years commenced from the *annus Passionis*, aligning with the *annus mundi* 5229 based on Eusebius' *Historia ecclesiastica.*
10 Jones, *The Victorian*; McCluskey, *Astronomies*, 86–87. Basically, the *Cursus* ended up with proposing a 'Latin' list of Easter dates coinciding with the Alexandrian tradition.
11 *Concilia Galliae*, 132; Ohashi, *The Easter Table.*
12 For Abbo's computus and his critique of the Dyonisian system, see Verbist, *Duelling with the Past*, 35–84.
13 Gennadius of Massilia, *De viris illustribus,* 88, pp. 107–108: "Victorius, natione Aquitanus, *calculator scripulorum*, invitatus a sancto Hilarius, urbis Romae episcopo, composuit paschalem cursum indagatione cautissima post quattuor priores qui composuerunt, i.e. Hippolytum, Eusebium, Teophilum et Prosperum, et pertendit annorum seriem usque ad annum quingentesimum trigesimum secundum, ita ut quingentesimo trigesimo tertio anno paschalis reincipiat sollemnitas eodem mense et die eademque luna, qua primum passio ac resurrectio Dominica facta est". My italics.
14 This system of measurement originated in ancient Greece and then developed within the Roman world: treatises by Varro, Vitruvius, Martianus Capella, Priscian, and Isidore of Seville are dedicated to this topic. See Hultsch, *Metrologicorum*, 108–128, pp. 49–124, vol. II. See also Chapter 6 in this book.
15 *In Calculo*, III, 37, p. 95. Beda, *De temporum ratione,* III, 275–278. Beda introduces also the notions of *momentum* ($\frac{1}{40}$ of one hour) and *minutum* ($\frac{1}{10}$ of one hour). A different discussion is offered by Hrabanus Maurus in his *De computo*, in which he explains how many *horae, puncti, minuta, partes, momenta, ostenta*, and *atomi* are contained within a week. Hrabanus Maurus, *De computo*, 27, 14–21, p. 205. Cf. Tannery, *Sur les subdivisions.*
16 Beda, *De temporum ratione*, IV, 278–280.
17 *Bedae opera de temporibus*, 333.
18 *Bedae opera de temporibus*, 170.
19 *Bedae opera de temporibus*, 206 and 218. See also Jones, *The 'lost'*, especially 213–219; Madan, *A Summary Catalogue*, 8837, vol. 3, 13.
20 Hagen, *Catalogus codicum Bernensium*, 285–286; cf. Mostert, *The Library*, BF134, p. 68. The letters indicating the manuscripts are consistent with those adopted in Peden's edition of Abbo's *Explanatio.*
21 Kinkelin, *Der Calculus*, 148.
22 Löwe, *Codices latini antiquiores, Supplement*, 19; Bischoff, *Mittelalterliche Studien*, 40.
23 Gugel, *Welche erhaltenen*, p. 51; Bischoff, *Manuscripts*, 125.
24 On Lupus of Ferrières, see Holtz, *L'humanisme*; Romano, *Lupo di Ferrères*; and Ricciardi, *L'Epistolario*, xiii–xv.

25 Lupus of Ferrières, *Epistulae*, 115bis, p. 111. This letter dates to 861.
26 Bischoff, *Manuscripts*, 127, n. 47.
27 Bischoff, *Manuscripts*, 143–145.
28 Lupus of Ferrières, *Epistulae,* 5, pp. 13–15.
29 Lupus of Ferrières, *Epistulae,* 5, p. 14.
30 Lupus of Ferrières, *Epistulae,* 5, p. 15.
31 Bischoff, *Manuscripts*, 125; Pellegrin, *Membra disiecta*, 5.
32 McCarthy, *The Arrival.*
33 Ohashi, *The Easter Table*, 144–147. In Oashi's opinion, the omission of Victorius' computus in Beda's *Historia ecclesiastica gentis Anglorum* might have been due to a sort of millenarianism supported but Victorius himself (147–149). Cf. Corning, *The Celtic*, 83–94.
34 Stahl, *Katalog*, 257–258.
35 It has been brought to modern scholars' attention by Warntjes, *A Newly Discovered Prologue.*
36 Columbanus, *Epistulae*, 2, p. 18.
37 Löwe, *Codices latini antiquiores, Supplement,* 19. Cf. Ferrari, *Spigolature.*
38 See Ritcher, *Bobbio*; and Zironi, *Il monastero.*
39 Mercati, *Le principali vicende.*
40 Riché, *Relations*, 13–18.
41 Leotebold's testament is edited in Prou and Vidier, *Recueil des chartes*; for the foundation of the monastery, see 5.
42 See Peden, "Introduction", xlviii–xlix.
43 *In Calculo*, III, 15, pp. 81–82: "Sin partium congregationem legeris, ut in quibusdam codicibus habetur, pro compositione congregationem accipe, quia congregatio multitudinis est coadunatio, conpositio vero non solum multitudinis sed etiam magnitudinis ad unum quiddam redacta continuatio".
44 Rose, *Verzeichnis*, 308–315; cf. Peden, "Introduction", xxxvii–xxxviii.
45 See d'Onofrio, "Introduction", xxiii–xxiv.
46 Dengler-Schreiber, *Scriptorium*, 207–210.
47 Van de Vyver, *Les oeuvres inédites*, 140. Sigebert of Gembloux, *De scriptoribus ecclesiasticis*, 139, PL 160, 578.
48 *Tabulae codicum*, 44–45; in this catalogue, the manuscript is dated from the 13th century. According to Guillaumin, the *De arithmetica* copied in it traces back to the 11th century; see Guillaumin, "Introduction", lxxxiv. Peden's opinion is that the *Explanatio* copied in this manuscript is from the 11th century, too; see Peden, "Introduction", xli. Van de Vyver links this codex to G, as they share Boethian texts; according to him, the codex comes from Gembloux; see Van de Vyver, *Les oeuvres inédites*, 140.
49 An edition of this text is in progress and will be hopefully delivered soon.
50 PL 90, 677–680.
51 Christ, "Uber das argumentum", the critical edition of the *Calculus* is on 132–136.
52 Friedlein, "Der Calculus"; Friedlein, "Victorii Calculus".
53 *Thesaurus novus anecdotorum*, 118–120.
54 PL, 139, 569–572.
55 Christ, "Uber das argumentum", the critical edition is on 136–152.
56 *Gerberti Opera mathematica*, 197–203.
57 *Gerberti oper mathematica*, 203. Cf. G, fol. 114v.
58 *Gerberti oper mathematica*, 203, n. 15.
59 See Chapter 6 in this book.
60 See Chapter 3 in this book.
61 Abbo, *Quaestiones grammaticales*, 50, 275.
62 We will come back to this in the last section of this chapter.

63 On this practice see Evans, "Introductions to Boethius'".
64 Evans and Peden, "Natural Science", 111–114.
65 See Chapter 5 § 5.2.1 in this book.
66 See above the description of Victorius' *Calculus*, specifically what I defined *Supplementary tables and texts* (*Olearia incipiunt pondera* and *Item mellaria incipienti*, in Peden's edition).
67 See Chapter 7 in this book.
68 *In Calculo*, I, 2, p. 65: "ad arithmeticam introductionis pontem construo".
69 *In Calculo*, II, 72, p. 116.
70 Isidore of Seville, *Etymologiae*, I, 30, 1 (Lindsay's edition): "Glossa graeca interpretatione linguae sortitur nomen. Hanc philosophi ad verbum dicunt, quia vocem illam de cuius requiritur uno et singulari verbo designat. Quid enim illud sit in uno verbo positum declarat".
71 William of Conches, *Glosae super Platonem*, 10, p. 67: "Unde commentum dicitur plurimum studio vel doctrina in mente habitorum in unum collectio. Et quamvis secundum hanc diffinitionem commentum possit dici quislibet liber, tamen non hodie vocamus commentum nisi alterius libri expositorium. Quod differt a glosa. Commentum enim, solam sententiam exequens, de continuatione vel expositione litterae nichil agit, glosa vero omnia illa exequitur. Unde dicitur glosa id est lingua: ita enim aperte debet exponere ac si lingua doctoris videatur docere".
72 On this, see Jeauneau, "Gloses et commentaires". On the development of the gloss as a literary genre, see in the same volume, the study by Hadot, "La préhistoire".
73 The literary practice of the *accessus ad auctores*, within its juridical, rhetorical, and philosophical frame, has been investigated by Quain, "Medieval accessus".
74 Boethius, *In Isagogen. Editio prima*, I, 1, 4–5 (hereafter *1 In Isagogen*).
75 Boethius, *In Isagogen. Editio secunda*, I, 5, 146–147 (hereafter *2 In Isagogen*); Boethius, *Commentarium in Perihermeneias, editio prima*, 31–34.
76 Boethius, *Commentaria in Categorias*, PL 64, 159A-161C.
77 John Scotus Eriugena, *Annotationes in Marcianum*, 3, ll. 1–23.
78 Remigius of Auxerre, *Commentum in Martianum*, *Accessus*, 65.
79 *In Calculo*, I, 3, p. 65: "Inpraesentiarum tamen intentio Victorii haec fuit, ut inerrato lector numerorum summas multiplicaret, divideret; seu proponeretur aliquid de artibus, quae numerorum ratione constant, ut arithmetica, geometria, musica et astronomia, seu quaestio inesset de mensura et pondere, quae omnia calculatori sunt curae. [...] Verum, quoniam omnia creata sunt in numero, mensura et pondere, in his singulis speculationem placet constituere, ut perspecta eorum natura, facilius pervideri possit singula quibus insunt <in> causa".
80 *In Calculo*, III, 41, p. 97: "More proemii solito mira brevitate captat auditoris benivolentiam, attentionem et docilitatem: benivolentiam, quod causa humilitatis non se calculandi argumentum, sed antiquos finxisse conmemorat; attentionem, quod se de divisione rei incorporeae corporeaeque pariter disputaturum denunciat, [...]; docilitatem, quod de tam magnis rebus locuturus, longitudinis fastidium vitans, conpendiosam divisionem lucide et aperte se agere promittit".
81 Cicero, *De inventione*, I, 20, pp. 75–76. Cf. Cicero, *Topica*, 97, p. 100.
82 Boethius, *Commentarii in Ciceronis Topica*, PL 64, 1042D-143A. Martianus Capella, *De nuptiis*, V, 545–546, p. 192. See Murphy, "Western Rhetoric".
83 The unity of this collection has been reconstructed by Pellegrin, "Membra disiecta", 13–16.
84 *In Calculo*, III, 41, p. 98: "Antiqui commenti sunt, quia philosophis placuit omnia infinita ac dubia redigere per probationem saltem ad verisimilitudinem, ut ex his aliquam speculationem sumerent, quia infinitorum scientiam, quae nulla est, conprehendere nequirent. Sed quod ait 'commenti sunt', id est finxerunt, videtur sonare eos non ipsam veritatis naturam tenuisse; quia fingere non solum

conponere verum etiam quandam vultus simulationem solet praetendere. Hoc tamen loco commenti sunt significat 'invenerunt' vel 'excogitaverunt', ut Terentius: 'Facite, fingite, invenite' [*Andria*, II, 334], licet fictio, id est simulatio, veritatis sit immaginatio. Unde commentatores eos vocamus, qui veritatem aliquo modo obscuris sententiis involutam multa verisimilia fingendo expositionis luce inluminant, quae inventa 'commentarios' vocant".

85 Caiazzo, "Abbon de Fleury", 23–24. Cf. Jeauneau, "L'usage", and Bezner, "Vela veritatis".

86 Cf. the definition of *integumentum* in the 12th-century *Commentum super sex libros Eneidos Virgilii* (attributed to Bernardus Silvestris), 3: "Integumentum est genus demonstrationis sub fabulosa narratione, veritatis involvens intellectum; unde et involucrum dicitur". For a detailed analysis of the use of the myth to clarify a text, see Chenu, "*Involucrum*".

87 Cicero, *De oratore*, I, 160, p. 159.

88 Macrobius, *Commentarii*, I, 2, 6–16, pp. 6–8 (hereafter *In Somn.*). See also Dronke, *Fabula*, 13–14.

89 The excerpt has been edited by Caiazzo, "Abbon de Fleury", 40, which is part of the section "Extraits des *Commentarii in Somnium Scipionis* dans le ms. de Paris, BnF, nouv. acq. lat. 1630, fol. 14v-16v et le ms. de Leyde, Bibl. der Rijksuniv., Vossius lat. F. 70 I, fol. 51r".

3 A Theological Premise

Number, Measure, and Weight

3.1 Augustine and *Wisdom* 11:21

To determine the background of the *Tractatus de numero, mensura et pondere*, let us turn first to Augustine's view of *Wisd.* 11:21.[1] Although Augustine (354–430) never wrote a specific treatise on the *Book of Wisdom*, he referenced the verse multiple times in works composed between 388 and 428, mentioning it 31 times.[2] In Augustine's early works, written before he had read the *Book of Wisdom*, there are references to the elements of the biblical triad, i.e., number, measure, and weight. These early references lay the groundwork for the conceptual framework later expounded in his exegesis of *Wisd.* 11:21. Augustine's early conception of numbers posits them as ontological constituents of the universe. Numbers, according to Augustine, are not independent and foundational aspects of reality properly speaking; rather, they form the basis of the world insofar as they represent divine rationality and wisdom.[3] This perspective is articulated in the *De ordine*, where Augustine emphasises how humans link sensible objects to mathematical concepts, demonstrating the immanence of numerical proportions in the cosmos. The *De musica*, too, contains a prefiguration of the sapiential triad, with Augustine arguing that the ontological structure of every created entity is determined by three elements: number, unity, and order, reflecting the ordered development of parts in both inanimate and animate entities.[4]

The first explicit occurrence of *Wisd.* 11:21 is found in his *De Genesi contra Manichaeos*.[5] In this context, Augustine employs the verse to illustrate that even the creation of seemingly useless or harmful animals is part of the overall harmony and order willed by God. Augustine emphasises that even withing the limbs of every living being, there exist measures and proportions contributing to the creature's harmonic structure and unity. These proportions ultimately depend on the supreme measure, number, and weight inherent in the perfect divine essence. Still with the aim of criticising the Manicheans, in the *Contra Faustum*, Augustine invokes the beauty of the created world as evidence of divine creation.[6] He finds this beauty in the

DOI: 10.4324/9781032643472-4

proportionate measures, numbers, and weights that play a role in forming the physical aspects of every creature.

In line with this, in the third book of the *De Trinitate*, Augustine incorporates number, measure, and weight into a broader discussion about *rationes causales*.[7] There, he portrays the work of creation as a progressive unfolding of measures, numbers, and weights impressed by God; for it is in fact in God that the numbers, measures, and weights of all things are found. Such a perspective stresses the divine (mathematical) imprint on the ordered and structured nature of the created world.

Augustine's discussion of the sapiential triad is complex, and the use of synonyms adds to the challenge of interpretation.[8] The semantic shifts in Augustine's terminology, thus, are crucial for understanding his view on this subject matter:

- *Species* and *forma*: these terms can refer to number in Augustine's framework. Number can represent the principle of form, defining the essence of an entity. It signifies what makes an entity what it is, allowing it to be known and defined by its specific difference.
- *Modus* and *unitas*: Augustine uses these terms in connection with measure. Measure can be seen as the divine operation that 'limits' beings, for God bestows a specific 'quantity of existence', determining the extent to which what exists can exist.
- *Ordo* and *pax*: Augustine associates these terms with weight. Weight can express the idea of general perfection and order. It signifies that each creature ontologically occupies its appropriate place. Additionally, it reflects the tension of all creatures towards their ultimate purpose or end.

In summary, God, for Augustine, is the ultimate source that 'limits', forms, and orders all creatures. Number, measure, and weight are intertwined concepts that capture the essence and orderliness inherent in the divine creation.

The most detailed and comprehensive exegesis of the *Wisd.* 11:21 by Augustine is found in the fourth book of his *De Genesi ad litteram*. In this work, when delving into the sixth day of creation, Augustine focuses on the significance of the number 6, which he describes as a 'perfect number' due to its being the sum of its factors (i.e., 1, 2, and 3).[9] In this instance, Augustine quotes *Wisd.* 11:21 to emphasise the order of creation: the verse serves to highlight the progressive realisation of harmony, size, and stability in creatures. Hence, Augustine attributes these qualities to the divine principles of number, measure, and weight found in God.

> But insofar as measure sets a limit to everything, and number gives everything its specific form, and weight draws everything to rest and stability, he [i.e., God] is the original, true and unique measure which defines for all things their bounds, the number which forms all things,

> the weight which guides all things; so are we to understand that by the words "You have arranged all things in measure and number and weight' nothing else was being said but 'You have arranged all things in yourself"?[10]

In this context, the sapiential triad is looked at in relation to the created world and God: number, measure, and weight – all and each of them – are God himself who provides all creatures with their specific properties, dimensions, order. Now, having assumed that God has arranged everything in himself according to number, measure and weight, Augustine engages with the question of whether these aspects exist prior to creation, either within God or outside him. Augustine considers both possibilities but seems to favour the idea that number, measure, and weight exist in God himself (they are *sicut ipse*), coinciding with the divine essence. From this perspective, God impresses upon creatures the mark of his own essence. This position, characterised as a 'metaphysics of intrinsic participation' by Aimé Solignac, suggests that these elements are inherent to God's nature, and creatures reflect the essence of the creator.[11] The alternative hypothesis, the 'metaphysics of equivocal causality', locates these elements externally to God and created things, and hence poses challenges. In this scenario, if number, measure, and weight are considered as 'thoughts' originating outside the divine intellect, creatures may no longer bear the mark of the creator but rather of some other principles (number, measure, and weight themselves). Although Augustine does not conclusively resolve this question, he introduces an analogy with colours to shed light on his perspective. Colours exist in coloured things, but in the mind of God, they are present as the *ratio colorum*, that is, a disposition that causes things to be coloured.[12] By analogy, the sapiential triad can be understood as an 'aspect' of divine rationality, according to which things possess number, measure, and weight.

In Augustine's exploration of how number, measure, and weight can be found in God and coincide with his essence, he advises moving beyond human experiential understanding of these terms. Instead of conceiving them as they appear in our everyday experiences, Augustine suggests considering them in their absolute and divine reality. In doing so, he employs a 'method of negation', asserting that God is beyond countable number, he is immeasurable measure, and unweighable weight.

> But this measure of spirits and minds is kept in bounds by another measure, this number formed by another number, and this weight drawn by the pull of another weight. But the measure without measure is the standard for what derives from it, while it does not itself derive from anything else; the number without number, by which all things are formed, is not formed itself; the weight without weight to which are drawn, in order to rest there, those whose rest is pure joy is not itself drawn to anything else beyond it.[13]

In summary, Augustine's interpretation of *Wisd.* 11:21 involves understanding number, measure, and weight in three general senses:

- Features of entities: number, measure, and weight are seen as characteristics of entities, serving as marks by which creatures resemble their creator.
- Metaphysical principles: number, measure, and weight are seen as coextensive with everything that proceeds from God, inasmuch as they are God himself.
- Conditions of creation: number, measure, and weight are viewed as principles according to which God brings everything into existence.

While the elements of the sapiential triad, especially when considered individually, can be linked to the Neopythagoreanism of young Augustine, the interpretation of the biblical verse is not inherently tied to arithmetical questions.[14] Rather, Augustine's exploration delves into the metaphysical and theological dimensions of these concepts within the context of creation and divine essence.

3.2 Claudianus Mamertus' *De statu animae*

The Augustinian hermeneutical strategy finds resonance in the works of Claudianus Mamertus. Of particular significance to us is Mamertus' *De statu animae* (composed around 470), which serves as the primary source for Abbo's *Tractatus de numero, mensura et pondere*. Mamertus dedicates three chapters in the second book of the *De statu animae* to elucidating the concepts of number, measure, and weight in relation to the body, the soul, and the Trinity.[15] It is exactly these three chapters that Abbo's 'theological premise' draws upon.

Let us provide first a historical and philosophical framework for Claudianus Mamertus, who is certainly not the most renown character of the earliest Middle Ages.[16] Born in the city of Vienne in the Lyonnais region, Mamertus lived during a time when the Burgundians settled in 443. A contemporary of Victorius of Aquitaine, Mamertus was a prominent figure in 5th-century Gaul, as attested by his mention in Gennadius of Marseille's *De viris illustribus*. Gennadius highlights Mamertus' rhetorical skills and his work *De statu animae*, which aimed to demonstrate that "something else, besides God, is incorporeal".[17] It is likely in Lyon that Mamertus encountered Sidonius Apollinaris (431–487), to whom the *De statu animae* is dedicated.[18] In an epistle to Mamertus, Sidonius commends his proficiency across various fields of knowledge, drawing comparisons to emblematic figures. According to Sidonius, Mamertus mastered the plectrum of Orpheus, the radium of Archimedes, delved into numbers like Chrysippus, and explored measures akin to Euclid. He is likened to other philosophical giants such as Pythagoras, Socrates, Plato, Aristotle, and Augustine.[19] Among these, the association with

Augustine is particularly noteworthy, considering that Mamertus' philosophy draws significant inspiration from Augustine's thought.[20]

Mamertus composed the *De statu animae* in the intellectual context of the 5th-century controversy surrounding the nature of the soul. This dispute, involving Mamertus and Faustus, the bishop of Riez, focused on the longstanding Christian Neoplatonic discussions about the relationship between the soul and the body and their modes of interaction.[21] The *De statu animae* was specifically written to counter Faustus' epistle *Quaeris a me*, where Faustus argued that the soul is determined by quantity, asserting its spatial inclusion in the body, that is, its extension within the space circumscribed by the limbs.[22]

While the final book of the *De statu animae* is dedicated precisely to refuting the arguments presented by Faustus in his epistle *Quaeris a me*, the first two pose the basis to address this controversy as they display Mamertus' viewpoint. In the first book, he defends the incorporeal nature of the soul based on philosophical reasonings. Mamertus presents some arguments to support ideas such as that God did not neglect to create something superior in ontological dignity to what is corporeal and local. There, he also contends that incorporeality is not synonymous with invisibility, spiritual substances cannot be perceived by sight, and they undergo neither increase nor decrease. Additionally, Mamertus argues that the soul performs a life-giving function in the body, and in this sense, the soul 'is in the body', while it is not determined by any quantity. The soul perceives corporeal things through the body whereas knows incorporeal things through itself.

Conversely, in the second book of *De statu animae*, Mamertus deploys a variety of philosophical authorities to support his stance. These sources include both pagan philosophers from the Greek and Latin traditions, the writings of the Church Fathers, and references from the Bible. Notably, Mamertus pays particular attention to Pythagorean works and authors, mentioning figures such as Philolaus of Croton, Archyta, Hippo of Rhegium, Aristaeus of Croton, and Diodorus of Aspendus, among others.[23] Moreover, the *De statu animae* proves to be an important source for what historians of philosophy may call the Plato Latinus, for it features two lengthy excerpts from Plato's *Phaedrus* (245c) and *Phaedo* (66b-67a), as well as a passage from a work attributed to Plato titled Περὶ φυσικῆς (which is, in fact, Apuleius' *De dogmate Platonis* I, 9). Mamertus cites various other titles of Platonic works, including the *Hipparchus*, *Laches*, *Protagoras*, *Symposium*, *Alcibiades*, *Gorgias*, *Crito*, and finally, the *Timaeus*. While this section of the work may appear as a broad and unsystematic review of Platonic psychology, its purpose is to uphold the account of the soul as autonomous from the body, rational, and positioned as something intermediate between 'what is simple and invisible' and 'what is divisible and visible'.

Mamertus devotes particular attention to *Wisd.* 11:21, considering it a crucial biblical authority that supports the thesis of the soul's incorporeality. Relying on this verse, he dedicates three chapters to elucidate how number, measure, and weight are applicable to the body, the soul, and God. Like Augustine did, Mamertus raises questions about the relationship between the sapiential triad and God himself. He grapples with the idea that if these three exemplars – number, measure, and weight – had been created by God, then it would follow that they had not been created according to number, measure, and weight. But this appears to contradict the scriptural assertion that *everything* was created based on these principles. Thus, Mamertus concludes that the exemplars must not have been created.[24] Furthermore, he posits that the sapiential triad served as a model in the creation of the world and consists in the measure by which we measure (*quo metitur*), the number by which we count (*quo numeratur*), and the weight by which we weigh (*quo penditur*).[25] As such, these exemplars are distinct from everything that is measured, numbered, and weighed, persisting even in the absence of numerable, measurable, and weighable bodies.[26]

Mamertus delves into the impact of number, measure, and weight on bodies, elucidating their role in determining the structure of constituent parts and the specific natural place these bodies occupy in the world.[27] While Mamertus draws inspiration from Augustine's ideas, his exposition differs from Augustine's approach to *Wisd.* 11:21. Mamertus – and Abbo, as we will see – expands on the effects of the three exemplars on corporality, grounding his explanations in a fundamental understanding of the qualitative physics of the elements. Like the body, the soul, too, is subject to measure, number, and weight. Its measure is envisaged as a capacity to house wisdom. The number of the soul becomes apparent when cardinal virtues are harmoniously expressed in human actions.[28] In accordance with Augustine's account, Mamertus attributes the weight of the soul to the human will and love: a sort of 'local weight' acts as a guiding force for the soul, drawing and directing it towards its most cherished love and desire: God.[29]

As in Augustine's *De Genesi ad litteram*, Mamertus engages in a dialectical negation of the sapiential triad to trace it back to God. God, then, is presented as measure without measure, number without number, and weight without weight.[30] As the Trinity, God is tied to number, measure, and weight in another way. For instance, weight is metaphorically expressed through the love of the Father towards the Son, which extends into human hearts through the Holy Spirit.[31] Thus, along with the sapiential triad, the Trinity itself, too, serves as a model (*specimen*) that extends its influence to bodies and souls. Bodies are considered vestiges (*vestigium*) of the Trinity as they exist individually, while also possessing a threefold nature, precisely being numerable, measurable, and weighable. Souls, in contrast, are likened to images (*imagines*) of the Trinity. They mirror the threefold nature of the Trinity as they operate based on the three faculties of memory, will, and understanding.[32]

As previously mentioned, Abbo's *Tractatus de numero, mensura et pondere* incorporates many ideas from Mamertus' *De statu animae*. While Abbo's 'theological premise' shifts away from addressing strictly psychological issues, it is worth remarking that Mamertus' treatise was a focal point for the discussion of philosophical problems precisely related to the soul. This was the case before and beyond Abbo's time – the treatise *De statu animae* losing its importance only in the 13th century.[33]

Mamertus gained prominence especially during the Carolingian period – an example being Cassiodorus' *De anima*, for which Mamertus' work represented a key source.[34] More generally, many concepts from the 5th-century controversy on the nature of the soul were revisited. One noteworthy notion from Mamertus that we find repetitively in the Carolingian period is that of locality (*localitas*), a term first introduced by Mamertus himself to convey the Greek ἀδιαστασία, that is, *inlocalitas*, which signifies not the absence of space but rather a property of what is 'extensionless' and not stretched into three dimensions.[35] In the *De ratione animae*, nowadays attributed to Pseudo-Hincmar, a set of questions addresses the relevance of the notion of locality: whether the soul is corporeal ("utrum anima corporea sit"), whether it is locally confined in the body ("utrum localiter teneatur in corpore"), and whether the soul moves with the body through places ("utrum anima cum corpore moveatur per loca").[36] The issue of the locality of a spiritual substance is also explored in Ratramnus' *De anima*: when contemplating whether the soul is in a place, Ratramnus asserts that the notion of locality is applicable solely to entities that can be divided into parts, a characteristic that is not applicable to the soul.[37] Even in Eriugena's *Periphyseon*, Mamertus' influence can be traced, at least at a linguistic level. Transitioning from the issue of the location of the soul to the metaphysical problem of the nature of space, Eriugena argues against the equivalence of quantity and space. In so doing, he employs precisely Mamertus' binomial *localitas-inlocalitas*.[38] Further evidence of the dissemination of Mamertus' psychological treatise in the mid-9th century can be found in a poem by Micon of Saint-Riquier expressing the desire to obtain a copy of the *De statu animae* to enhance the one owned by his monastery.[39] Additionally, Lupus of Ferrières, in an epistle dated 859 to Odon of Corbie, requests the works of Faustus of Riez, the opponent of Mamertus within the 5th-century controversy on the soul.[40]

Now, Faustus' epistle *Quaeris a me* is transcribed alongside the *De statu animae* in manuscripts such as Paris, BnF, lat. 2164, a codex that also allows us to retrace the path of Mamertus' treatise towards the abbey of Fleury. This manuscript, analysed by Michel Huglo, contains Mamertus' work and Calcidius' translation and commentary on the *Timaeus*. Huglo suggests that it originated from Fleury's scriptorium, being copied in Fleury from another manuscript coming from Corbie, likely in the 10th century.[41] Considering Mamertus' influence on the Carolingian debate, it is indeed likely that that a copy of the *De statu animae* was present in Corbie, especially if one considers that Ratramnus, author of the *De animae*, served as monk there. One of the arguments supporting the Fleurisian origin of the manuscript Paris, BnF, lat. 2164 is precisely the substantial use Abbo makes of the *De statu animae* in the 'theological premise' of his *Expositio*, namely, in his *Tractatus de numero, mensura et pondere*.

To the best of my knowledge, Abbo is the author who demonstrated the most attention to Mamertus' psychological treatise. In contrast to the Carolingian texts mentioned earlier, which mainly relied on Mamertus' idea of 'non-locality' of the soul, Abbo's *Tractatus de numero, mensura et pondere* extensively relies on Mamertus' account of number, measure, and weight.[42] It is noteworthy, however, that unlike Calcidius and Macrobius, explicitly mentioned in the *Explanatio*, Mamertus and his work are never directly referred to by Abbo. This could be explained by supposing that Abbo considered Mamertus a well-known author who did not require explicit mention; or alternatively, that Abbo's colleagues at Fleury were already acquainted with the *De statu animae*. On the other hand, it is also possible that the absence of explicit quotations is due to the limited authority granted to Mamertus by 10th-century authors, especially those interested in speculative arithmetic and calculus. This is particularly relevant given the type of text in which Abbo's borrowings from Mamertus occur – a commentary on Victorius' multiplication tables, serving also as an introduction to arithmetic. However, as already pointed out, the borrowings from Mamertus' *De statu animae* primarily concentrate in the *Tractatus de numero, mensura et pondere*, functioning as a theological prelude to the actual arithmetical commentary.[43]

3.3 Arithmetical Readings of *Wisdom* 11:21: Hrabanus Maurus and John Scotus Eriugena

Although not direct sources for Abbo's *Tractatus de numero, mensura et pondere*, Hrabanus Maurus and John Scotus Eriugena provide insights into the intertwining of *Wisd.* 11:21 with arithmetic – specifically Boethian number theory. Their interpretations, while distinct from Augustine's, offer a theoretical background that aligns with the overall philosophical framework of Abbo's *Explanatio*. In drawing parallels between these thinkers and Abbo, it becomes apparent that Abbo's decision to emphasise the sapiential triad in his prologue aligns with the belief in an intrinsic correlation between the biblical verse on creation and the rationality embedded in arithmetic.

Hrabanus Maurus was a prominent figure in the Carolingian Renaissance. Born around 780 and associated with key figures like Alcuin of York and Lupus of Ferrières, Hrabanus became the abbot of Fulda in 822 and later served as the bishop of Mainz from 847 until his death in 856.[44] Known to modern scholarship for his extensive literary output, Hrabanus Maurus' works include didactic treatises such as *De arte grammatica*, *De computo*, and *De rerum naturis*, as well as commentaries on the Scripture. To him we owe the first medieval systematic commentary on the *Book of Wisdom*.[45] Regarding *Wisd.* 11:21, Hrabanus interprets the verse as referring to the equal justice of God, emphasising God's impartiality as reflected in the structure of every creature. To support this interpretation, Hrabanus cites biblical passages (*Psalms* 147 and *Job* 28) and notably draws upon a philosophical authority. Although he doesn't explicitly mention the source, it is clear that he extensively quotes Boethius' *De arithmetica* in this context, as we can read from the passage below.

> BUT YOU HAVE DISPOSED EVERYTHING BY MEASURE AND NUMBER AND WEIGHT. But you disposed everything according to number, measure and weight. Because [God] does everything according to truth, judgement and justice, and there is nothing inordinate and inappropriate in his works and actions, but he regulates and arranges all things moderating his judgements with the balance of his perfect equity. This statement can also be applied to the status of every creature: because the omnipotent creator created according to a specific number, measure and weight all his works, which he made in the beginning. [...] Hence, someone else states: "All things, whatever they were, which were created at the beginning seem to be shaped by reason of numbers, since this was the original exemplar idea in the mind of the creator. From it was derived the multiplicity of the four elements; from it were derived the changes of the seasons; from it the movement of the stars and the turning of the heavens. Since things are like this and since the states of all things are shaped by the binding together of numbers, it is necessary that number in its own substance maintains itself evenly at all times, permanently". Who, therefore, has the capacity to understand this measure, this number, and this weight of God, so as to be able to expound his making up of all things, within these three definitions? I think that the quality relates to the measure, the quantity to the number, and the substance to the weight.[46]

In Hrabanus' interpretation of *Wisd.* 11:21, the quotation from Boethius' *De arithmetica* plays a central role. This philosophical authority reinforces what appears to be a Neopythagorean reading of the biblical verse. According to this interpretation, God, in creating the world, imparted order to creatures, and this order is fundamentally rooted in mathematical rationality. The Boethian passage is crucial for elucidating why and how mathematical rationality serves as the foundation of the universe. Hrabanus' exegesis suggests that, in creating the world, God established an order grounded in mathematical principles. As per Boethius, number possesses a stable essence and serves as the model that God employed to structure reality. The ordered nature of reality, then, is intimately connected to numbers, which exist in an intermediary state between the created world and God. Within this Boethian metaphysical framework, Hrabanus endeavours to discern the correspondences of number, measure, and weight in reality. In his explanation, these three aspects represent distinctive facets of the constitution of all creatures. Specifically, number relates to quantity, measure to quality, and weight to substance.

In Hrabanus' exegesis, the Boethian, Neopythagorean account of reality, while not extensively explored or used as the foundation for his own views, holds significance. Hrabanus' *Commentary on the Book of Wisdom* represents the first medieval exegesis of this biblical verse, and in this context, Boethius' *De arithmetica* is the sole philosophical source referenced. Although Hrabanus does not delve deeply into developing his philosophical perspective on numbers and the world based on the Boethian, Neopythagorean account, the juxtaposition of these two texts initiates an interpretative and philosophical trend that gains prominence in later medieval thought.

This trend is further advanced and expanded upon by another Carolingian scholar, John Scotus Eriugena (810 ca. – *post* 877), translator from Greek, theologian, and philosopher associated with the *schola palatina* of Charles the Bald. A more extensive and profound exploration of *Wisd.* 11:21 and Boethian number theory is evident in his *Periphyseon*, which is a philosophical dialogue between a master and a disciple unfolding in five books.[47] The philosophical exploration of the two characters commences with an investigation into the ultimate genus that encompasses the entire ontological realm, identified as 'nature' – this nature including both 'things that are' and 'things that are not'. Eriugena employs a dialectical division based on the concept of creation to understand this nature, resulting in the identification of four species: (1) the nature that is not created but creates; (2) the nature that is created and also creates; (3) the nature that is created but does not create; and (4) the nature that neither is created nor creates. These four species coincide to the ontological levels of reality, that are, respectively: God, the primordial causes, the concrete physical world, and God conceived as the final cause. This fourfold ontological scheme describes the procession (*processus*) of creatures from and their return (*reditus*) to the 'uncreated nature', namely, God.

The third book of the *Periphyseon* is of great interest for us since it introduces an arithmetical excursus in which the quotation of *Wisd.* 11:21 occurs twice, accompanied by extensive references to Boethius' *De arithmetica*.[48] The arithmetical excursus delves into a metaphysical dilemma concerning the status of the creatures and God, and is motivated by the idea that creatures are both created and eternal. According to one character in the dialogue, i.e., the master, God encompasses the ideas or primal causes and the reasons of all things, and simultaneously, God is present in each created thing. From this point of view, creatures can be seen as both eternal, when considered as enfolded in God, and made when viewed within the physical manifestation of the world – spatially and temporally located.[49] Thus, the potential risk of a pantheistic interpretation arises, where God might appear to be present in all creatures. This prompts the disciple to seek clarification. The reply of the master involves a sharp question that only apparently eludes the topic at stake: "Are you versed in the art of arithmetic?"[50] This query hints at the forthcoming arithmetical discussion that aims to provide a solution to the metaphysical dilemma and, by extension, to the exegesis of *Wisd.* 11:21.

To grasp Eriugena's reasoning on God, creatures, and numbers, it is useful to briefly review the key features of Boethius' metaphysics and number theory. First, Boethius posited that numbers are ontologically prior to created things, existing in their own substance without undergoing any change. Second, numbers play a fundamental role in structuring the world as the primary exemplar used by God in the act of creation. Finally, unity is presented as the cause for all numbers.[51] Eriugena incorporates these Boethian principles into his discussion, aiming to illustrate the relationship between God and creatures especially through the analogy with unity and numbers. Let us look at how Eriugena utilises Boethius' number theory and *Wisd.* 11:21 to arrive at his philosophical conclusion about God and creatures.

In response to the master's question about his familiarity with the art of arithmetic, the disciple demonstrates his proficiency in the subject. He provides a definition of arithmetic, which can be traced back to Boethius' *De Trinitate*, and according to which arithmetic is the science that deals with numbers used in counting.[52] It specifically explores intellectual numbers, that is, those apprehended by the intellect alone and not associated with any physical subject. These intellectual numbers are invisible and incorporeal.[53]

Eriugena, like Hrabanus before him, relates the discipline of arithmetic to the created physical realm. He justifies this connection by invoking a thesis articulated by Boethius in his *De arithmetica*, stating that arithmetic, as the foremost among mathematical sciences, pertains to nature. This is because the substances of all things, both visible and invisible, adhere to numerical principles examined by arithmetic itself. Eriugena further reinforces this philosophical doctrine by referencing the authority of Scripture, particularly *Wisd.* 11:21, and drawing a connection to the renowned pagan philosopher Pythagoras.[54]

> For not only does it [arithmetic] subsist as the immutable basis and primordial cause and principle of the other three branches of mathematics, namely, geometry, music, astronomy, but also the infinite multitude of all thing visible and indivisible assumes its substance according to the rules of numbers which arithmetic contemplates, as the supreme philosopher Pythagoras, the first inventor of this art, testifies when he gives good reasons for asserting that the intellectual numbers are the substances of all things visible and invisible. Nor does the Holy Scripture deny this, for it says that "all things have been made in measure and number and weight".[55]

While the first quotation of *Wisd.* 11:21 is introduced to stress the role played by numbers in configuring the substance of created things, the second one (reported below) relates to the issue of the ontological priority of numbers, drawing on a specific aspect of Boethius' number theory, namely, the properties of perfect numbers.[56] Perfect numbers possess the specific arithmetical property of being equal to the sum of their factors – an example being the number 6, which is equal to the sum of 1, 2, and 3. The notion of perfect numbers is then linked by Eriugena to the number of days in the creation narrative. He asserts that the number 6 is not perfect because God created the world in six days; rather, God created the world in six days because the number 6 is perfect. This arithmetical value, reflecting perfection, also holds some ontological significance. Eriugena's overarching aim is to demonstrate the ontological primacy of all numbers over created reality, included place and time. Since God created everything in accordance with number, as indicated in *Wisd.* 11:21, all created things – whether subject to time or not, along with time itself – are ontologically posterior to numbers.

> It is then credible, or likely that this most mighty and divine exemplar [i.e., the number 6] in which God made his works had a temporal beginning, when in it not only the things which are in times but also the times themselves and the things which subsist beyond the times were constituted by the creator of all things? Therefore, no man of sound wisdom would have any doubt about the eternity of the numbers if he made use of the argument concerning the number 6 only, for what is understood about its eternity must similarly be understood of the perenniality of the others. For not of the number 6 alone, but generally of the totality of all the numbers was it said, "God made all things in measure and number and weight". But if places and times are counted among all the things which God made, the intellectual numbers subsisting in their science alone necessarily precede the places and times in the perpetuity of their nature [...].[57]

In Eriugena's viewpoint, what he calls intellectual numbers exist as the object of a science, that is, within the intellect. Despite this, their creation is not attributed to the human intellect, but rather to God, who established them eternally within the unity or monad.[58] This stance is crucial to illuminate the analogy between the relationship involving God and creatures and that one involving unity and numbers. The extensive arithmetical and arithmological digression in the third book of the *Periphyseon* is aimed at exploring the parallelism between these relationships. Eriugena's articulation of the unity's role in relation to numbers draws on influences from Pseudo-Dionysius' *Divine Names*, Maximus the Confessor's *Ambigua*, and notably, Boethius' conception of unity and number theory.[59] According to Eriugena, the unity, or monad, serves as the beginning, middle, and end of all numbers.[60] This implies that the unity acts as their cause, common factor, and the equality ratio in which inequality ratios can be reduced through specific arithmetical rules. Eriugena explicitly references his source for this doctrine as Boethius' *De arithmetica*, which he identifies as *de mathesi libri*.

> All numbers subsist eternally in the monad and while they flow forth from it, they do not cease to be in it since they cannot abandon their natural state. For whether by multiplication or by division they proceed from it and return to it in accordance with the rules of the art which considers their reasons. [...] What I shall say of the marvellous and divine constitution and proportion of the superparticulars and the superpartiens and of the multiple superparticulars and the multiple superpartiens, which the species receive individually from the unity? [...] If anyone diligently wishes to know of these things let him carefully read the books of the great Boethius on mathematics.[61]

Eriugena's idea is that numbers flow forth from unity and return to it, with the understanding that inequality ratios can be reversed back to equality ratios.

> Also, if the numbers flow forth from the monad as from some inexhaustible source and, however much they are multiplied, come to an end in it, they would surely not be flowing forth from it before their flowing forth they had not subsisted in it as in their cause; nor would they seek their end in it if they did not know by their natural motion that there were not eternally abiding in it their causes towards which they never desist from returning through the same stages by which they were flowed forth from it by the rules of analysis by which every inequality is recalled to equality. Now the rules of analysis will be found at the beginning of the second treatise on mathematics of the great Boethius by every student who pursues the marvellous investigation of such natures.[62]

Eriugena's exploration of the eternity and causality of unity in relation to numbers provides a foundation for understanding the analogy between God and creatures. By emphasising the eternal nature of unity as the principle of the infinite progression of numbers, Eriugena draws parallels between the timeless origin of unity and the divine nature. Unity, as the cause of numbers, contains them potentially and eternally, reflecting God's eternal and causal relationship with the created world. Numbers are above time and so is their origin, unity. They are infinite, thus their origin cannot be finite, since an infinite progression cannot stem from a finite beginning.[63] This conceptualisation extends to creatures, which, like numbers, are considered eternal in their potential existence within God. The act of creation represents their unfolding from this eternal potentiality into actualised, concrete existence.[64] This framework allows Eriugena to articulate a metaphysical understanding of the relationship between the eternal, unchanging nature of God (analogous to unity) and the temporal, evolving nature of the created world (analogous to the numerical progression).

Similar to Hrabanus, Eriugena interweaves the authority of the Scripture with that of Boethius. However, in contrast to Hrabanus, for whom Boethius' *De arithmetica* merely reinforces the interpretation of the Scripture, Eriugena invokes *Wisd.* 11:21 and Boethius' work to develop his unique perspective on the ontological status of numbers and to address a metaphysical dilemma concerning the existence of creatures in relation to God. Both biblical and mathematical sources are integrated into a broader, original philosophical discourse.[65] This distinctive approach is also evident in Abbo's work. The 'theological premise' built around the terms referenced in *Wisd.* 11:21 serves as an introduction to essential aspects of Boethian number theory and calculus.

3.4 Abbo's *Tractatus de numero, mensura et pondere*

The *Tractatus de numero, mensura et pondere* is designed to elucidate the exemplary causes (specifically, number, measure, and weight) according to which God structured the world. This initial theological discussion stems from what Abbo considers the subject matter of arithmetic and calculus: these disciplines provide the means to analyse and divide all temporal and spatial magnitudes, making them applicable to every quantitatively determined entity. Through the study of arithmetic and calculus, then, Abbo believes that one can identify the exemplary number, measure, and weight among and within created things. In essence, Abbo envisions arithmetic and calculus as disciplines that facilitate the recognition of the intrinsic mathematical rationality proper to God, as alluded to in *Wisd.* 11:21. This, according to Abbo, represents the loftiest goal of these disciplines.

The exploration of number, measure, and weight intertwines theology and philosophy, insofar as the examination of the three exemplars falls within a specific philosophical domain, namely, physics. To comprehend this outcome, one must consider that Abbo's analysis is situated within a broader discourse on wisdom. According to the etymological understanding dating back to antiquity, philosophy is the love of wisdom. Now, Abbo emphasises that the path to wisdom involves the pursuit of knowledge across the seven liberal arts. This is stressed by Abbo through a metaphor deeply ingrained in Western philosophical tradition: the genuine abode of wisdom, to which philosophy guides, is supported by seven pillars akin to those adorning Solomon's temple.[66] However, it is Boethius' *De arithmetica* that fundamentally angles Abbo's account on wisdom: it is characterised as the "subtle contemplation of divine things, perfect cognition of immutable things, and full understanding of truth".[67] The seven liberal arts, in the end, lead humans to what is immutable and true.

Within this examination of wisdom, Abbo introduces a novel interpretation of Boethius' *De hebdomadibus*, offering insights into Abbo's own understanding of free will and its role in humans' choice to embark on the path of wisdom, embodied in the study of the seven liberal arts. According to Abbo, wisdom is not solely confined to the cognitive realm, but it also influences human behaviour by enabling individuals to distinguish between good and evil, and it directs human actions aiming to restore the divine image within humans.[68] In line with Boethius' *De hebdomadibus*, Abbo asserts that created things are not substantial good in and of themselves.[69] However, his departure from Boethius is evident in the rationale behind this assertion. Abbo contends that what is essential to something never lacks in the thing to which it is essential – akin to fire never being cold, and the sky never being motionless. In contrast, goodness is treated differently. Abbo argues that goodness is not inherent to human beings because if it were, there would be no room for free will. Every human choice and action would be dictated by the necessity of an innate virtue ("ex ingenitae virtutis necessitate"). In other words, this would lead to ethical determinism, a concept that Abbo finds inconceivable. The impossibility of sin, in Abbo's perspective, is exclusive to God, the supreme good. Human beings, on the other hand, are meant to be guided by wisdom, indicating to them which goods to pursue, including philosophy.[70] Wisdom itself serves as the remedy to ethical determinism, so that the capacity to sin is the counterpart of the capacity to make mistakes.[71]

Within this framework, the love of wisdom, that is, philosophy is somehow mirrored by the powers of the soul. Abbo aligns with the philosophical tradition represented by Calcidius, Macrobius, and Boethius, acknowledging the tripartite division of the powers of the soul.[72] According to this schema, all living beings possess the vegetative power (*vis crescendi*), some also have the vegetative and sensitive powers (*vis sentiendi*), and only a select few have the vegetative, sensitive, and rational powers (*vis discernendi*). These powers are deemed essential for initiating the journey towards wisdom, guided by the love of wisdom, namely, by philosophy. The philosophical pilgrimage commences with the knowledge of visible things, which are explored through the senses. It gradually progresses to encompass the understanding of invisible and divine realities, ultimately reaching the profound mystery of Trinitarian unity.

> The love of wisdom seems to imitate, in a certain way, the triple power of the soul, which gives the power to grow to some of the living beings, to grow as well as perceive to others, and to grow, perceive, and discern to some others. Although it happens that these degrees are found there [i.e., in the soul], yet it is not wisdom but the love of wisdom that, growing within us, incidentally, advances from accidents, when it rises from visible things through invisible things to the ineffable unity of the Trinity.[73]

After observing that wisdom is attained though philosophy, which in turn proceeds via the liberal arts, and that the human soul predisposes human beings to such a theoretical pilgrimage, Abbo proceeds to outline the actual path of philosophy. Thus, he delineates philosophy into three distinct branches: ethics, logic, and physics.[74] Each branch serves a specific purpose in the pursuit of wisdom. Ethics encompasses the study of religious morality and human conduct and is further divided into four cardinal virtues (prudence, justice, fortitude, and temperance), which provide a framework for individuals in leading a virtuous life. Based on Boethius' *De topicis differentiis*, logic is defined as the systematic analysis of discourse (*ratio disserendi*) and is divided into two main parts: a topical part (*pars inveniendi*), concerned with the discovery of arguments and ideas; and an analytical part (*pars iudicandi*), focusing on judging and evaluating arguments. Moreover, logic comprises the arts of the trivium (grammar, rhetoric, and dialectic) and plays a crucial role in upholding rigor and truth in philosophical discourse: it safeguards against falsehoods and ensures that philosophical judgments are grounded in valid and reliable reasoning.[75] Physics, on its side, is illustrated as follows:

> Physics is a part of it [i.e., philosophy] in which the faculty primarily concerned with numbers, measures, and weight is refined – we also intend to pursue this aspect under the guidance of Victorius, if the right conditions will be met, through the fourfold mathematical disciplines of the quadrivium, whose example the formation of the whole world required. However, [...] it should be known that in contemplating, not in taking actions, [the physics] itself examines these things. In this way, the pure contemplative activity of the soul leads to [the knowledge of] other causes of things that exist within nature.[76]

In light of this quotation, Abbo's conceptualisation of physics emphasises its mathematical character and its foundational role in comprehending the structure of created reality. Physics comprises the four disciplines of the quadrivium – arithmetic, music, geometry, and astronomy – and thus deals with number, measure, and weight. The emphasis on the quadrivium in physics underscores the interconnectedness of mathematical knowledge and the causes shaping the order of creation. From this perspective, Abbo's commentary on Victorius' *Calculus*, a mathematical treatise, becomes a valuable resource for exploring the intricacies of this numerical rationality and its implications for understanding the structure of the created reality.

The causes operative within the realm of nature ("naturae beneficio"), which constitute the focus of physics, encompass the concepts of number, measure, and weight utilised by God in the act of world creation.

The three fundamental, mathematical principles, according to Abbo, could not have been created. In line with Mamertus' exposition, Abbo contends that should number, measure, and weight be subject to creation, they would lack their own exemplar causes, thereby contravening the authoritative scriptural dictum of *Wisd.* 11:21.[77] Rejecting the prospect of their creation, Abbo, akin to Augustine in *De Genesi ad litteram*, contemplates the whereabouts of number, measure, and weight preceding the act of creation.[78] While Augustine's inquiry delves into the internal or external placement of these principles in relation to God, Abbo directs his attention towards distinguishing them from created entities, concluding that they do not coincide with quantified bodies.[79] For Abbo, the crux lies in discerning such mathematical exemplars from both quantity conceived as an accident, on one hand, and any concrete quantified item occupying space, on the other.[80] Number, measure, and weight stand apart from everything that can be numbered, measured, and weighed, analogous to the distinction between greatness and what is great, equality and what is equal, and beauty and what is beautiful.[81] Number, measure, and weight serve as the conditions for created entities to be numerable, measurable, and weighable. However, in order to apprehend them *per se* (i.e., as absolute and not determined by any material conditions), Abbo employs the dialectical approach previously advanced in the aforementioned Augustinian exegetical work. The comprehension of these principles necessitates subjecting them to a negation: in contrast to created, finite bodies, the three exemplars emerge as unnumberable, unmeasurable, and unweighable. Indeed, every created entity, including the mass of the world itself as well as all the bodies therein, is finite and possesses quantifiable attributes. Beyond the confines of the world, however, there exists nothing that is quantitatively determined. Yet, the potential for quantifiability persists. For example, outside the finite mass of the world, any measurable place is absent, but the condition for local measurement still stands. Even in the absence of place or body, thus, the conditions for quantifiability endure. The underlying cause that enables place and body to define themselves as quantified and finite entities remain intact.

> Hence, one must grasp the sign of that weightless weight, immeasurable measure, and innumerable number, which are all and each coeternal and individual, always and everywhere the one God. For all body, by which and which we count, weigh, or measure, are countable, weighable, and measurable. [...] Indeed, anything finite has a definite size. Therefore, the entire mass of the universe, being compacted by finite bodies, since one body serves as the end for another, undoubtedly is itself finite, and thus measurable. However, beyond the universe, since there is no place, there is no local measure where there is no place; and when what can be measured is lacking, that by which it can be measured remains, because the immeasurable measure does not lack if there is no finite measurable thing. In this way, the principle of weight persists even when there is no longer what can be weighed, and number does not end when what is countable ends.[82]

Number, measure, and weight constitute the conditions for being enumerated, measured, and weighed, persisting even in the absence of tangible, quantified entities. Abbo thus designates them as *quo numeratur*, *quo metitur*, and *quo penditur* – terms denoting that on the basis of which counting, measuring, and weighing is possible. In this context, counting, measuring, and weighing do not denote the actions of quantitatively defining something according to specific units of measurement. Instead, they signify the existence of things according to quantitative determinations. What, then, is the specific causal action that these exemplars undertake? Let us turn to Abbo's own words for clarification.

> Therefore, any part of the earth, whether large or small, like a tiny pebble or the minutest speck of dust that cannot even be like a visible point, has both a measure for its size and a number for the arrangement of its parts, by which the higher parts are distant from the lower, the right from the left, the front from the back. Thus, even the smallest part can be divided into two, for it is a body. Its weight is observed in this way: when this tiny part is lifted from its origin and released in the upper element, namely water, it does not rest with an incessant motion until, by its natural weight, it reaches the ground. Similarly, you shall follow this path of reasoning with regard to other bodies, for whether it is a droplet of water, a particle of air, or a spark of fire, it is carried upward or downward by its weight, [has size] by measure, and [occupies] space by number.[83]

The configuration and arrangement of each individual entity, as well as the ordered structure of the entire world, are determined by number, measure, and weight. They operate to shape reality in the following manner. Number defines the proportion of the parts of each entity, both externally and internally. This implies that number influences the spatial relationship with other entities and the arrangement of the parts within it – within an Aristotelian context, one could refer to this influence of number as determining the 'spatial differences' of something.[84] Measure imparts a formal and quantitative 'limit' to all entities based on the specific characteristics proper to each entity. Finally, weight governs the natural motion of each entity, which, in turn, determines the specific place for each of the four elements and bodies in general. Every particle of air or drop of water, even the smallest bodily thing – even the unperceivable one – is subject to number, measure, and weight as long as they, however minuscule, possess parts, occupy space, and have motion.[85]

In addition to number, measure, and weight, two more divine features are discernible among bodies as well as in rational souls: unity and ternary. Drawing primarily from Mamertus (*De statu animae*, II, 5–6), Abbo contends that both bodies and rational souls exhibit traces of the God as Trinity. This is why everything, whether corporeal or incorporeal, is characterised as *unum* and *trifarium*.[86] Bodies are singular yet at the same time are somehow three, inasmuch as they are numerable, measurable, and weighable. Conversely, rational souls are also singular but operate in accordance with three psychological faculties. As an incorporeal entity, the soul can apprehend number, measure, and weight in their essence, and through them, it can contemplate and assess anything that is numerable, measurable, and weighable – while it is through the body that it perceives quantities that fall under these categorisations.[87]

Abbo's commentary on *Calculus* is constructed upon these foundations. Once the mathematical rationality underlying the world is comprehended, one can embark on the arithmetical journey towards wisdom. Along this disciplinary path, physics will expound on the theological assumptions, revealing a mathematical ratio underpinning the world. Encompassing arithmetic and calculus, physics investigates every realm of reality susceptible to measurement. Measurement and division, thus, are means to elucidate the nature of all entities, both corporeal and incorporeal. From this perspective, calculus corresponds to a science of reality and is deeply intertwined with physics itself.

The *Tractatus de numero, mensura et pondere* concludes with a reference to unity, the supreme divine principle. It is in the opening lines of the *Explanatio* that Abbo delves into the meaning of unity, explicating the initial sentence of Victorius' *Calculus*: "unitas illa unde omnium multitudo numerorum procedit". As we will explore in the following chapters, the unity-composition binomial represents a central theme in Abbo's philosophy.

Notes

1 Augustine's view of *Wisd.* 11:21 has been assessed by Peri, "*Omnia mensura*"; Cf. La Bonnardière, "Le livre de la Sagesse".

2 A synoptic table of Augustine's works in which the biblical verse occurs is in La Bonnardière, *Biblia augustiniana*, 295–297. It is worth specifying that Augustine relies on the *vetus Latina* Bible, which has "omnia *in* mensura et numero et pondere disposuisti", instead of Jerome's *vulgata*, in which the preposition 'in' is missing. The *vulgata* includes the version of *Wisd.* 11:21 which is present in the so-called *Itala* version, one of the Latin translations of the Bible constituting the *Vetus Latina* and dating back to the 2nd century AD. The *Italia* version reads: "omnia mensura et numero et pondere disposuisti". See *Vetus latina*, 449–452. Cf. the entry *Sagesse (livre de la)*, in *Dictionnaire de théologie catholique*, vol. XIV/1, col. 703.

3 Augustine, *De ordine*, II, 15, 43, pp. 130–131; Augustine, *De ordine*, II, 14, 39, p. 129; and Augustine, *De ordine*, II, 19, 49, p. 134. In line with this, Augustine traces back the indefectibility, the eternity, and the validity of mathematical concepts to God, the supreme divine unity. Furthermore, he proves the immortality of the human soul by the presence of mathematical truths in the human mind, as we can read in Augustine, *De immortalitate animae*, IV, 5–6, pp. 106–107. As David Albertson pointed out, these aspects of Augustine's thought can be labelled as Neopythagorean. See Albertson, *Mathematical Theologies*, 68–73. On Augustine's attitude towards Pythagoras, see Feichtinger, "Nothing Rash", and Janby, "Christ and Pythagoras".

4 Augustine, *De musica*, VI, 17, 57, pp. 231–232.
5 Augustine, *De Genesi contra Manichaeos*, I, 16, 26, pp. 92–94.
6 Augustine, *Contra Faustum manichaeum*, XXI, 6, p. 575.
7 Augustine, *De Trinitate*, III, 9, 16, p. 143.
8 For a systematic overview of these semantic shifts and the polysemous meanings of the sapiential triad throughout all Augustine's works, see Roche, "Measure"; Beierwaltes, "Augustins Interpretation"; Harrison, "Measure"; Pizzani, "Qualche osservazione", 304–315; Bettetini, *La misura delle cose*, 127–147 and 18, for a synoptic table.
9 Augustine, *De Genesi ad litteram*, IV, 2, 6, pp. 94–95. See also Augustine, *De civitate Dei*, XI, 30, p. 350, where Augustine addresses the numerical structure of the universe, which is created in accordance with the number 6. In this case, *Wisd.* 11:21 is presented as a praise to God, who knows all numbers. The juxtaposition of *Wisd.* 11:21 and the number of the days of creation is a traditional feature within the exegesis of the *Genesis*. It can be found also in Isidore, *Etymologiae*, III, 4, 1, p. 11: "In multis enim sanctarum scripturarum locis quantum mysterium habent elucet. Non enim frustra in laudibus Dei dictum est: 'Omnia in mensura et numero et pondere fecisti'. Senarius namque [numerus] qui partibus suis perfectus est, perfectionem mundi quadam numeri [sui] significatione declarat".
10 Augustine, *De Genesi ad litteram*, IV, 3, 7, p. 99. Translation from Augustine, *The Literal Meaning*, 246.
11 Cf. the notes in Augustine, *La Genèse*, 635–639.
12 Augustine, *De Genesi ad litteram*, IV, 5, 11, p. 101.
13 Augustine, *De Genesi ad litteram*, IV, 4, 8, p. 100. Translation from Augustine, *The Literal Meaning*, 247 (slightly modified).
14 For traces of Neopythagoreanism in young Augustine, see Albertson, *Mathematical Theologies*, 68–73.
15 Claudianus Mamertus, *De statu animae*, II, 4–6, pp. 111–119. Hereafter simply *De status animae*. The chapters at stake are titled: *De mensura, numero et pondere corporis*; *De mensura, numero et pondere animae*; *De mensura, numero et pondere divinae Trinitatis quo modo haec tria ipsa Trinitatis sit*.
16 See Courcelle, *Les lettres*, 221–222, where attention is brought to the upswing of literary studies in the Gaul region in the last quarter of the 5th century; for Mamertus, see especially 223–235.
17 Gennadius of Massilia, *De viris illustribus*, 84, p. 90: "Claudianus, Viennensis ecclesiae presbyter, vir ad loquendum artifex et ad disputandum subtilis, composuit tres quasi *De statu vel substantia animae* libros, in quibus agit intentione, quatenus ostendat esse aliquid incorporeum praeter Deum".
18 The dedicatory letter allows us to determine the date of the composition of Mamertus' treatise, as Mamertus addresses Sidonius using the title of patrician, which Sidonius obtained before becoming bishop of Auvergne in 471; *De statu animae*, *Praefatio*, 18–20. On Sidonius Apollinaris, cf. Courcelle, *Les lettres grecques en Occident*, 235–240.
19 Sidonius Apollinaris, *Epistolae*, IV, 3, 5, pp. 72–74.
20 Lambert, "Mamertus Claudianus", 1357–1358; Bömer, *Der lateinische Neuplatonismus*, 36; and Courcelle, *Les lettres*, 230–232.
21 I refer to Fortin, *Christianisme*; and Di Marco, *La polemica*.
22 Faustus of Riez, *Epistula (Quaeris a me)*, 10–11.
23 *De statu animae*, II, 7, pp. 120–122.
24 *De statu animae*, II, 4, p. 111. Cf. *In Calculo*, II, 10–11, p. 69.
25 *De statu animae*, II, 4, pp. 111–112. Cf. *In Calculo*, II, 12, pp. 69–70.
26 *De statu animae*, II, 4, p. 113. Cf. *In Calculo*, II, 13, p. 70.
27 *De statu animae*, II, 4, p. 114. Cf. *In Calculo*, II, 15, p. 71.
28 *De statu animae*, II, 5, p. 117.

29 *De statu animae*, II, 5, pp. 117–118. Cf. *In Calculo*, II, 16, p. 72, where Abbo paraphrases Mamertus' passage.
30 *De statu animae*, II, 6, p. 118.
31 *De statu animae*, II, 6, pp. 118–119. Abbo's *Tractatus* does not present this remark.
32 *De statu animae*, II, 6, p. 119. Cf. *In Calculo*, II, 16, p. 72.
33 Boethius can be considered an exception, as he refers to Mamertus' work not to discuss about psychology. In his *De institutione musica*, he relied on some examples on hearing and voice presented in the *De statu animae* to elaborate his own concept of *musica humana*. This has been pointed out by Restani, "Le radici". Boethius' treatise on music would be, thus, the first case of the use of Mamertus' text in a context far from psychology. As for Mamertus' influence on medieval psychology, it is worth noting that in the 12th century, Nicholas of Montiéramey and John of Salisbury still refer to *De statu animae*. On this, see Chenu, "Platon à Citeaux", 102. A detailed inquiry of the medieval treatises *de anima* is offered by Tolomio, *L'anima*; and Tolomio, "L'origine".
34 Cassiodorus, *De anima*, 49–50. Carolingian discussions regarding the nature of the soul were prompted by an alleged questionnaire sent by Emperor Charles the Bald to Hincmar of Reims (806 ca. – 882) and Ratramnus of Corbie (d. 870 ca.). Mathon has reconstructed this questionnaire by comparing and analysing the questions addressed in the following texts: *De ratione animae* by pseudo-Hincmar, *De anima* by Ratramnus, *De diversis quaestionibus* by Gottschalk of Orbais. A synoptic table comparing relevant passages of these texts is in Mathon, *L'anthropologie*, 269. It is worth noting that in the 9th century, the psychological debate resurfaced also within discussions on angelic bodies and the existence of eternal fire as a place of torment, as addressed by John Scotus Eriugena. John Scotus Eriugena, *De praedestinatione liber*, XVII, 7–8, pp. 182–186. I consider the theories discussed in Eriugena's *De praedestinatione* as part of the Carolingian debate about the soul only with respect to the specific issue concerning the quantity and locality of the soul therein addressed. See Mainoldi's introduction to the critical edition of Eriugena's *De praedestinatione*, especially ix–xli.
35 Mathon, "L'anthropologie" 283–287. The influence of the *De statu animae* within the Carolingian controversy about the soul has been analysed by Cristiani, "L'espace". See also Bömer, *Der lateinische Neuplatonismus*, 111–128.
36 Pseudo-Hincmar of Reims, *De ratione animae*, PL 125, 932–937.
37 Ratramnus of Corbie, *De anima*, 211 for the definition of *localitas*.
38 I follow Cristiani's interpretation offered in "L'espace", 150–160. Cf. John Scotus Eriugena, *Periphyseon*, III, 731B, 161.
39 Micon of Saint-Riquier, *Carmina*, 102, 335.
40 Lupus of Ferrières, *Epistulae*, 111, 106.
41 Huglo, "D'Helisachar". In previous studies, Michel Huglo has also shown the Fleurisian origin of two Parisian manuscripts: BnF, lat. 2164 and BnF, lat. 1747, which transmit respectively Macrobius' *Commentary on the Dream of Scipio* and Ambrosius' *De fide Sanctae Trinitatis*. See Huglo, "Trois livres"; and Huglo, "La réception". These manuscripts are not included in Mostert, *The Library*. On the glosses to *De statu animae*, Calcidius' *Commentary on the Timaeus*, as well as on similarities between the diagrams in Paris, BnF, lat. 2164 and Berlin, Staatsbibliothek, Phill. 1833 (referred to as F in the previous chapter), see Caiazzo, "Abbon de Fleury", 17–19.
42 More precisely, extracts of the *De statu animae* can be found in Abbo, *In Calculo*, II, 12–16, pp. 69–72.
43 The *Tractatus de numero, mensura et pondere* takes the whole chapter 2 in Peden's edition of Abbo's *Explanatio* (*In Calculo*, II, 1–16, pp. 65–72). It is worth reiterating that the *Tractatus* is in no way transmitted independently from the *Explanatio* within the manuscript tradition.

44 On Hrabanus see Depreux, Lebecq, Perrin, and Szerwiniack, *Raban Maur*.
45 An essential tool to study Hrabanus's reading of the Scripture is Cantelli Berarducci, *Hrabani Mauri*. See also Leonardi, "Aspects"; and Savigni, "L'interpretazione". Glossing the biblical text, Hrabanus presents diverse issues concerning Christianity, such as the conduct of the clergymen, the unity of the Christendom, the problem of the heretics and the 'false Christians'. Hrabanus' exegetical approach involves juxtaposing passages from the Bible and the writings of the Church Fathers to aid readers in understanding the Scripture.
46 Hrabanus Maurus, *Commentariorum in librum Sapientiae libri tres*, PL 109, 723 B-C: "*Sed omnia in mensura, numero et pondere disposuisti.* Quia omnia secundum veritatem, judicium et justiciam facit, nec est aliquid in operibus ejus, et in factis inordinatum seu inconveniens, sed omnia judicia sua summae aequitatis libra moderando temperat atque disponit. Potest et haec sententia ad creaturarum omnium conditionem transferri: quia creator omnipotens omnia opera sua quae ab initio condidit, certo numero, mensura et pondere creavit. [...] Hinc et per quemdam dicitur: 'Omnia quaecunque a primaeva rerum constructa sunt, numerorum videntur ratione formata, hoc enim fuit principale in animo conditoris exemplar. Hinc enim quatuor elementorum multitudo mutata est, hinc temporum vices, hinc motus astrorum coelique conversio. Quae cum ita sint, cumque omnium status numerorum colligatione fingantur, eum quoque numerum necesse est in propria semper sese habentem aequaliter substantia permanere'. Quis ergo hanc mensuram, hunc numerum et hoc pondus Dei sufficit comprehendere, ut constitutiones ejus universas in his tribus definitionibus valeat enarrare? In mensura puto quod constet qualitas, in numero quantitas, in pondere ratio".
47 Of the vast literature consecrated to Eriugena's thought, I mention two of the most recent and systematic studies, namely, Otten and Allen, *Eriugena and Creation.*; and Guiu, *A Companion*. The former gathers most of the current bibliography on Eriugena divided by year of publication starting from 2000 and up to 2014. Among the classical studies, see Cappuyns, *Jean Scot Érigène*; and Jeauneau, *Études Érigéniennes*.
48 Crucial for the present discussion are the following studies: Jeauneau, "Jean Scot", and Duchez, "Jean Scot".
49 John Scotus Eriugena, *Periphyseon*, III, 642A-646C, 35–41. Hereafter *Periphyseon*.
50 *Periphyseon*, 650C-651A, 47.
51 *Arithm*. I, 1–2, pp. 6–12. Cf. Crialesi, "Numbers".
52 *Periphyseon*, 651B, 48. Cf. Aristotle, *Physics*, IV, 219b 5–9; and Boethius, *De Trinitate*, III, 171, ll. 132–134 (hereafter *Trin.*): "Numerus enim duplex est, unus quidem quo numeramus, alter vero qui in rebus numerabilibus constat". For the presence of Boethius' theological treatises in the *Periphyseon*, see d'Onofrio, "Giovanni Scoto".
53 *Periphyseon*, 651B-651C, 48.
54 For the reference to Pythagoras in this passage of Eriugena's *Periphyseon* see also Panti, "Pythagoras", 70–71.
55 *Periphyseon*, 651D-652A, 49. Translation by Sheldon and Williams, in John Scotus Eriugena, *Periphyseon*, III, 103. Hereafter referred to simply as Sheldon-Williams, followed by page number.
56 *Periphyseon*, III, 655D-656A, 54. Cf. *Arithm*., I, 19, 9–20, 1, p. 42.
57 *Periphyseon*, III, 656B-656C, 54–55. Sheldon-Williams, 113.
58 *Periphyseon*, III, 658B, 57.
59 Cf. Ps.-Dionysius's *Divine Names* in Eriugena's Latin translation in PL, 122, 1149B; and Eriugena's translation of Maximus the Confessor, *Ambigua* 6, 41, p. 98. In the *Periphyseon*, Eriugena offers another mathematical example

expressing the analogy between God and creatures, namely, that of the radii of a circle which are contained in its centre, that is, the point. See, for instance, *Periphyseon*, III, 639D, 32.

60 *Periphyseon*, III, 652B, 49.

61 *Periphyseon*, III, 653D-655A, 51–53; Sheldon-Williams, 109. Cf. *Arithm.*, I, 32, 2, p. 67: "Every type of inequality arises from a prior equality so that equality is itself as it were the matrix and, taking the force of a root, it gives depth to the types and orders of inequality" (Masi, *Boethian Number Theory*, 114). For the arithmetical procedure that allows to produce different inequality ratios starting from the equality ratio, see *Arithm.*, I, 32, 3–28, pp. 67–73. Different ways of formalising this procedure in modern arithmetical terms have been developed, see for instance Otisk, "The Interpretations", 37–40.

62 *Periphyseon*, III, 655B, 53; Shendon-Williams, 109-111. Cf. *Arithm.*, II, 1, 1, p. 78: "From the very source of equality and inequality we may see every species come forth, and we may resolve every species of inequality into equality as thought to a certain element of its own origin" (Masi, *Boethian Number Theory*, 122). The rules for reducing the inequality ratios to the equality ratio are explained in *Arithm.*, II, 1, 2–9, pp. 78–81, and are more extensively discussed in Chapter 5.

63 *Periphyseon*, III, 652C-D, 50.

64 *Periphyseon*, III, 657B, 56.

65 The epistemological implications about numbers, namely, how they are grasped by the human intellect and are related to the perception of material things are investigated by O'Meara, "The Metaphysical Use", 142–148, in which it is remarked that Eriugena's conception of numbers and Proclus' theory of mathematics share common features.

66 For instance, Cassiodorus and Alcuin of York refer to Salomon's temple when talking about the liberal arts. See Cassiodorus, *Institutiones*, II, *Praefatio*, 89; Alcuin of York, *Disputatio de vera philosophia (De grammatica)*, *Prologus*, PL 101, 853. On this, see d'Alverny, "La Sagesse".

67 *In Calculo*, II, 1, pp. 65–66. Cf. *Arithm.*, I, 1, 1, p. 6: "Est enim sapientia rerum quae sunt suique immutabilem substantiam sortiuntur comprehensio veritatis"; *Arithm.*, I, 1, 5, p. 7: "Est enim sapientia earum rerum quae vere sunt cognitio et integra comprehensio".

68 *In Calculo*, II, 2, p. 66.

69 *In Calculo*, II, 3, pp. 66–67.

70 *In Calculo*, II, 4, p. 67. Cf. Boethius, *De hebdomadibus*, 191–192, ll. 130–143 Hereafter referenced as *Hebd.*

71 Abbo's exploration of ethical determinism and wisdom can be intriguingly connected to his treatment of opinion and contingency in *Letter* 10, addressed to an anonymous bishop (PL 139, 433C). In this letter, Abbo discusses the nature of oaths, considering them as instances of spoken assertions that can be verified based on current and past events. While it is easy to verify whether a deed *has been* or *is being* realised, oaths that involve promises of future deeds cannot be easily verified by human reason. Abbo suggests that oaths regarding future events may deceive us due to contingent causes, unless they pertain to something happening out of necessity. In this context, opinion serves as the epistemological mode associated with contingency and oaths. The choice of what is considered good or bad, leading to wise action, is seen as subject to contingency and, consequently, is based on opinion. This insight into the contingency of knowing is paralleled by the contingency of events and actions. In this respect, ethical determinism cannot be an option. Special thanks to Christophe Grellard for drawing my attention to this connection.

72 Cf. Calcidius, *In Platonis*, 229–235, pp. 244–248 (hereafter *In Tim.*); *In Somn.*, I, 14, 10, p. 78; Boethius, *2 In Isag.*, I, 1, p. 136. Abbo seems to rely more on Macrobius, as Irene Caiazzo points out Caiazzo, "Abbon de Fleury", 22, especially n. 33.

73 *In Calculo*, II, 7, p. 68: "Amor sapientiae videtur quodammodo imitari triplicem vim animae, quae animatorum aliis solum crescendi, aliis crescendi pariter ac sentiendi, quibusdam etiam crescendi, sentiendi ac discernendi effectum attribuit. Quos gradus etsi illic invenire contingit, tamen non sapientia sed amor sapientiae in nobis meliorando ex accidenti proficit, dum a visibilibus per invisibilia ad inenarrabilem Trinitatis unitatem consurgit".

74 It is worth pointing out that on fol. 183v of manuscript Bern, Burgerbibliothek, B 56, which comes from Fleury, we find a diagram concerning the parts of philosophy, to which Abbo could have subscribed. Mostert specifies that the diagram resembles the one found in a text attributed to Alcuin (PL 101, 945–950). Mostert, *The Political Theology*, 164–166, where he analyses the 'virtues' reported in the diagram concerning the division of philosophy. Cf. also the diagram illustrating Ramsey Abbey's *curriculum* in manuscripts Oxford, St John College, 17, fol. 7r and London, British Library, Harley 3667, fol. 6v. According to Cyril Hart, this diagram stems from Abbo's teaching at the English abbey. Hart, *Learning and Culture*, 93–100.

75 *In Calculo*, II, 8 and 10, pp. 68–69.

76 *In Calculo*, II, 9, p. 69: "Est autem una pars eius [*i.e.* philosophiae] phisica (*sic*), qua praecipue numeri, mensurae et ponderis continetur excogitata facultas, quam etiam duce Victorio persequi deliberamus, si erit otium, per quattuor matheseos disciplinarum quadruvium, quod et eorum exemplo indiguerit totius mundi formatio. At vero, [...] sciendum quod non agendo sed contemplando haec ipsa discutit, quemadmodum alias rerum causas, quae vigent naturae beneficio, pura tradit animi speculatio".

77 *In Calculo*, II, 10, p. 69. Cf. *De statu animae*, II, 4, p. 111.

78 Augustine, *De Genesi ad litteram*, IV, 3, 7, p. 99.

79 According to Abbo, quantifiability is one of the main features that characterise bodies. Corporality and quantity are tightly interlaced in Abbo's thought, to the point that wherever a body is, quantity must be there too. On this, see Chapter 5, especially § 5.

80 *In Calculo*, II, 11, p. 69.

81 *In Calculo*, II, 12, pp. 69–70.

82 *In Calculo*, II, 13, p. 70: "Hinc capias oportet indicium illius non pensi ponderis et inmensurabilis mensurae et innumerabilis numeri, quae tria simul aequiterna semper individua ubique et ubicumque tota unus Deus sunt. Omnia etenim corpora, quibus quidque vel numeramus vel pendimus vel metimur, et numerabilia sunt et ponderabilia et mensurabilia. [...] Habet enim certum magnitudinis modum quicquid finale est. Proinde mundi moles universa, quoniam ex finitis est conpacta corporibus, quippe cum alterum corpus alteri finem faciat, procul dubio ipsa finalis est, propter quod et mensurabilis. *Extra mundum autem, quoniam locus esse non potest, localis mensura, ubi locus nullus est, non accedit, et cum desit quod metiri possit, quo possit metiri superest, quia finito mensurabili, inmensurabilis mensura non deficit.* Eatenus equidem ratio ponderis, eo quod pendi potest cessante, non perit et numerus finito numerabili non finitur". Italics is mine and is meant to point out what Abbo paraphrases from Mamertus. Cf. *De statu animae*, II, 4, pp. 112–113.

83 *In Calculo*, II, 15, p. 71: "Quaelibet ergo terrae pars magna vel minima, ut lapillus exiguus vel minutissimus pulvisculus, qui puncto visibili contingi non quaeat, habet et mensuram pro modulo sui et numerum pro ratione partium, quibus distant superiora ab inferioribus, dextra a sinistris, priora a posterioribus, unde etiam quamlibet minimum dividi in duo potest, quia corpus est. Pondus vero illius hinc animadvertitur, quod hoc ipsum minimum, ab origine sua sublatum et in aqua superiore videlicet elemento dimissum, motu infatigabili non quiescit, donec

naturali pondere ad solum terrae pervenerit. Hunc ad modum rationis tramitem in reliquis tene corporibus, quia vel aquae guttula *vel aeris particula vel ignis scintillula pro sui modulo mensura, pro loci spatio numero, pro naturali motu sursum aut deorsum pondere rapitur*". Italics is mine and is meant to highlight the part of the text in which Abbo distances himself from Mamertus. Cf. *De statu animae*, II, 4, pp. 113–114.

84 On the Aristotelian notion of spatial differences, see Crialesi, "Absolute Spatial Differences".

85 What is also worth noticing here is Abbo's use of some specific jargon that betrays a certain naturalistic sensibility (e.g., *lapillus*, *pulvisculus*, *aeris particula*). In his *De statu animae*, Mamertus lacks, for example, a comprehensive and careful terminology specifying the smallest parts of all four physical elements.

86 On the term *trifarium*, see *De statu animae*, II, 6, p. 119.

87 *In Calculo Victorii*, II, 16, p. 72; *In Calculo*, II, 14, p. 71. Cf. *De statu animae*, II, 6, p. 119.

4 Henology

4.1 Neopythagorean Elements: Unity and the Flow of Numbers

Specific features of ancient and medieval Neopythagoreanism revolve around four foundational points.[1]

1. Mathematics is seen as a comprehensive set of disciplines related to number, leading to an understanding of all levels of existence. Number, in this context, is revealing of both the physical and divine realms.
2. Mathematical entities constitute an intermediate level of being, for they can be placed between the divine and what is subject to change. Their significance consists also in bridging different levels of existence.
3. The divine is conceptualised as numerically one (monad) and as transcendent and ineffable (henade). This we may call henology.
4. Number is viewed as revealing the fundamental nature of beings and the structural order of the universe. In this sense, we may speak of a 'numerical ontology'.

These points definitely apply to Abbo as well. As emerged from the previous chapter, Abbo would subscribe to points (1), (2), and (4). Regarding point (3), we can look at two key doctrinal elements that underpin Abbo's theory of unity: firstly, the comparison or connection of God to the arithmetical unity, and secondly, the link between the way created beings depend on God and the way numbers spring from unity. In the case of Abbo, one may add on to these elements and bring the notion of goodness into the picture. God as arithmetically one, that is, the One, is also the ultimate good from which creatures intended as individual goods stem. This derivation of singular goods from the supreme one relies on an arithmetical model insofar as it mirrors the flow of numbers from unity. Such an outlook can be traced back to the philosophical lineage of Middle Platonism and to some ancient thinkers such as Eudorus of Alexandria and Moderatus of Gades.[2]

The idea of number as a flow of quantities originating from the monad and composed of other monads is to be found in Nicomachus of Gerasa's *Introduction to Arithmetic*.[3] Whereas it is in his *Theologoumena arithmeticae* that

DOI: 10.4324/9781032643472-5

a pronounced parallelism between the numerical unity and the divine one is elaborated. There, the numeric unity is conceived as capable of preserving the units constituting numerical multiplicity while remaining identical to itself – in this acting like its divine counterpart, which preserves the existence of multiple things while remaining transcendent. The equation between God and the arithmetical unity becomes more compelling as the latter is seen as something giving rise to all, being generated by nothing but itself, and thus existing without a beginning or an end. It does not seem that Nicomachus points to a subordination of one of these principles to the other, but rather that he emphasises the comprehensive potentiality and self-sufficiency inherent in both the divine and the numeric ones, suggesting a more profound analogy between the two.[4]

In his works on arithmetic and music, drawing chiefly from the writings of Nicomachus, Boethius similarly conceives the arithmetical one as the source of all numbers. The arithmetical unity is explicitly termed the 'mother' (*mater*) of the entire mathematical plurality, as both even and odd numbers proceed from it like in a continuous flow (*profluunt*), which consists of units.[5] This generative power of the arithmetical unity is particularly evident in the case of the so-called prime and 'non-composite' numbers (i.e., 3, 5 or 7). Specifically, in these instances, unity is also identified as the 'common measure', being the sole factor by which these numbers can be divided and, significantly, the only factor capable of generating them through multiplication.[6] Furthermore, the equality ratio (i.e., 1:1) is explicitly recognised as the principle from which other inequality ratios linking three different numbers (e.g., 4, 6, and 8) flow.[7]

Unlike Nicomachus, though, Boethius does not explicitly juxtapose the arithmetical and the divine ones in his mathematical works like the *De arithmetica* and the *De institutione musica*. While not overtly referencing Christian theology in these writings, the concept of mathematical unity is somewhat implicated in the theological discourse found in his *Opuscula sacra*.[8] Certain passages in these texts, particularly the *De hebdomadibus*, suggest a subtle reference to arithmetical unity when describing God as the supreme good and creatures as goods.[9] Let us see how. The central concern of the treatise is how created substances can be deemed substantial good simply by virtue of their being. Boethius, through a rigorous concatenation of inferences, concludes that substances are good not on account on their essence, but because they originate from one original goodness.[10] From a different perspective, one could posit that Boethius contemplates how the supreme good, a unique and simple principle, unfolds within the context of creaturely plurality, which bears God's features of being (*esse*) and being good (*esse bonum*). This dynamic is precisely articulated through the vocabulary of 'flow' (*fluere*) – a term that is used to describe the derivation of numbers from unity and the inequality ratio from the equality ratio in both the *De arithmetica* and the *De institutione musica*. In this framework, individual beings are considered good because they flow from the highest supreme good. They possess intrinsic goodness as they draw their being from the first principle, which is unique and simple, wherein 'being' and 'being good' coincide.[11] The supreme good facilitates the flow of being (*esse*) and, consequently, the being good (*esse bonum*) into the myriad created substances.

> Those [substances], since are not simple nor could they be at all unless that which alone is good willed them to exist, and since their existence *flowed* from the will of the good, they are said to be good. For the first good, since it is, is good in that it is; but the second good, since it *flowed* from that whose very being is good, is also good itself. But the very being of all things *flowed* from that which is the first good and which is so good that it rightly may be said to be good in that it is.[12]

The carrying of both goodness and being from the simplest principle to creatures is expressed through the terminology of the arithmetical unity initiating a flow of units and giving rise to all numbers. I argue it is from this perspective that Neopythagorean instances regarding both numerical and divine unity can be discerned in the *De hebdomadibus*: the supreme good flows into the multifaceted realm of created beings, endowing them with goodness and existence recalling the arithmetical unity engendering every number via a flow. Both of these flows illustrate how the unity (be it the numerical or the divine one) unfolds within multiplicity.

Further – yet feebler – echoes of such Neopythagorean elements can also be discerned in the *De Trinitate*. In the second book, after delineating an object and a specific method for each of the three speculative branches of philosophy – namely, physics, mathematics, and theology – Boethius describes what theology is. According to him, theological inquiry is dedicated to investigating the only form that is immune to the mutability of matter, namely God. The divine form is thus characterised as *unum* precisely because it lacks matter and movement, deriving its 'what it is' (*id quod est*) from nothing else. In other words, it does not require further formal determination from outside itself, as it is self-contained and consists solely in its own essence. Everything apart from God derives its being and formal determination from various elements. In contrast, God lacks this plurality of determinations and consequently, number within him, allowing him to be not only truly one but also simply unity ("unitas tantum").

> But the divine substance without matter is form and thus it is one, and it is 'what it is' (*id quod est*): for the other things are not 'what they are' (*id quod sunt*). For everything has its being (*esse*) from the things from which it is, namely, from its parts, and it is 'this' and 'that', namely, its parts conjoined, but not 'this' or 'this' separately – as

> when a human being consists of soul and body, and she is [both] soul and body, not the body of soul separately, [...]. But that which is not from 'this' and 'that', but only is 'this', that truly is 'what it is'; and it is the most beautiful and powerful because it relies on nothing. Therefore, this truly is one, in which there is no number, nothing in it except 'what it is'. [...] But God indeed differs from no God, [...]. But where there is no difference, there is absolutely no plurality, hence no number; therefore, only unity.[13]

Inheriting Nicomachus' view, Boethius seems to have transposed certain Neopythagorean ideas within a Christian theological framework. It is primarily through Boethius that these Neopythagorean doctrines found their way into the Latin Middle Ages. An illustration of this connection between the creator and creatures expressed in Neopythagorean terms can be found in the works of John Scotus Eriugena.[14] In a passage from his *Periphyseon*, that I have examined in the previous chapter, Eriugena elucidates how creatures perish in the physical world and yet simultaneously possess eternity in God. He introduces an arithmetical type of reasoning that presupposes the parallelism between God and the arithmetic unity, drawing connections between numbers and creatures.[15] According to Eriugena, numbers are eternal insofar as they are contained in their cause, flowing from unity, which serves as an inexhaustible source. The use of the verb 'to flow' (*fluere*), as discussed above, may distinguish Neopythagorean speculations, indicating, on the one hand, the derivation of numbers from unity and, on the other hand, the dependence of the multitude of creatures on the creator. This specific terminology of 'flow' along with the analogy between God and unity are also adopted by Abbo, as we will explore further below.

4.2 The Highest Good, *individuum*, and Unity

Abbo can be indeed considered part of this Neopythagorean tradition I just sketched, in as much as he both talks of the divine unity in arithmetical terms and describes all derivation from the (divine and numerical) unity in terms of 'flowing'. To analyse these issues, let us retrace Abbo's reasoning starting from the end of the *Tractatus de numero, mensura et pondere*, where Victorius' statement about unity is interpreted. In his *Calculus*, Victorius claims: "the unity from which the multitude of numbers proceeds, which is proper to arithmetic, does not admit any kind of division since it is simple and does not consist of any composition of parts".[16] According to Abbo, the unity Victorius refers to is God. Abbo regards God as (1) the highest good, (2) the *individuum* which is the beginning and end of everything, and (3) the arithmetic unity. Let us focus on these three features and highlight how Abbo relates different Boethian theories: the one concerning the simplicity of God and the goodness of created beings (from the *De hebdomadibus* and the *De Trinitate*); the one concerning division and universals (from the *De divisione* and the *Commentaries on the Isagoge*); and the number theory (from the *De arithmetica* and the *De institutione musica*, combined with Calcidius' *Commentary on Plato's Timaeus* and Macrobius' *Commentary on the Dream of Scipio*).

Abbo aligns with the general perspective found in Boethius' *De hebdomadibus*: God is the primary cause of goodness, upon whom and from whom all the goods pertaining to creatures, as well as creatures themselves inasmuch they are good, depend and flow.[17] This conception of God based on the *De hedbomadibus* is then related to the one exposed in the *De Trinitate*, according to which God is pure form and lacks all matter.[18] For Abbo, then, God imparts form to things while remaining immutable and simple. While creatures differ due to the conferred form, no change occurs in God.

> Since there is a principal simple cause of all existing things, whatever good proper to creatures, while proceeding from it, remains substantially in it. However, we do not profess that the origin of the highest good, from which and through which and in which all things proceed and exist, is material. Rather, we say that all things are good through participation in it. For the divine substance has always been and is without matter, although it does not lack form. By impressing form on individual things like a seal, it undoubtedly makes them differ from itself without any change in itself by increase or decrease.[19]

The reference to forms conferred by God to creatures prompts Abbo to specify the nature of such forms and how they proceed from God's simplicity to the plurality of creatures. In delineating this process, he grounds his reasoning on Boethius' account of universals, also anticipating some philosophical developments concerning forms as 'singular' that will fully emerge only later in the 12th century, through Gilbert of Poitiers.[20]

When God confers a form, he provides each creature with a specific difference that suits the nature of that particular being. This difference corresponds to a form, and many individuals can be related to it through some process of abstraction. Abbo argues that universals, such as genus and species, originate from individuals alone, as it is at the level of individuals that some shared traits, termed by Abbo as *dividua*, can be distinguished. *Dividua*, thus, represent those properties found within individuals that can be recognised as shared.[21] This 'dividuality' results from the similarity among individuals: *dividua*, for instance, can be that which renders a human being similar to other human beings or other animals. Through the collection of such *dividua* shared by several individuals (*collectio similitudinis*), it becomes possible, according to Abbo, to consider the shared features that enable the understanding of universals and formulate theorems specific to each discipline.

> And this difference pertains to the nature of each individual, for many individuals are subject to the forms that Plato calls 'ideas'. Where there is plurality, there is also the quantity of numbers. For from individuals arise *dividua*, from which universals are made, and through which the collection of similarities occurs, leading to the formation of species and theorems through each art.[22]

The acquisition of knowledge about universals (and forms, broadly speaking) appears to be grounded not in an Augustinian illuminationist or intuitionist framework, but rather in a genuinely Aristotelian *a posteriori* model. The crucial aspect in grasping this concept lies in the comparison of universals and theorems. The reference to theorems can be comprehended in the context of Calcidius' treatment of intellection, which is divided into knowledge and discursive reasoning. Calcidius characterises the former as pertaining to things perceptible by wisdom alone, such as God and ideas or forms, while the latter relates to things connected to technical precepts, like theorems.[23] In contrast to Calcidius, Abbo groups forms and theorems together, apparently placing them on the side of discursive reasoning. In this case, the epistemic model Abbo has in mind is clearly that of inductive abstraction, as explained in Boethius' *Second Commentary on the Isagoge*. Let us consider two examples to understand how the *collectio similitudinis* works in the case of collecting parts from an individual, or more precisely, when shared features pertaining to human beings are considered. Rationality can be considered a trait (a *dividuum*, Abbo would say) of the individual human being, which is found in all other human beings. One can identify rationality in each human being and collect this shared feature, and from this operation the species 'human being' can be obtained. Similarly, extension can be considered a trait of all material things. The notion of quantity, fundamental in formulating theorems in various disciplines, can be obtained by identifying and collecting this feature shared by every extended thing. As explained by Boethius in his *Second Commentary on the Isagoge*, it is possible to conceptually grasp the species starting from different individual things (x_1, x_2, x_3, ..., x_n), isolating the shared feature (x), which is their species, and then thinking of it alone, independently from other features. To grasp the species 'human being', it will then be necessary to abstract the nature of 'humanity' from different real humans and gather it within what we may call the "similitudo humanitatis".[24]

Keeping in mind the way Abbo conceives of forms, that is, *dividua*, it is worthwhile now to look at his notion of *individuum*. Building on the three meanings of 'individual' exposed in the *Second Commentary on the Isagoge*, Abbo explains that 'individual' can be defined:

1 Metaphysically, as that which lacks parts, such as God, the soul, and unity.
2 Physically, as that which cannot be divided, like a diamond.
3 Logically, as that which cannot be predicated of others, such as 'Socrates', which cannot be predicated of any other human beings nor can the parts of which the concrete subject Socrates is composed of be predicated of it.[25]

The metaphysical sense of the notion of individual reveals another aspect of Abbo's understanding of the divine principle, specifically God as *individuum*. Individuals must be situated within a broad ontological setting, in which what is complex derives from what is simpler, as well as what is composite derives from what lacks parts. According to Abbo, reality unfolds from an absolutely simple principle, and he explicitly states that: "undoubtedly, there is a certain individual (*individuum*) who is both the end and the beginning of everything", identifying this *individuum* as God.[26] While Boethius, in his *Second Commentary on the Isagoge*, presented units and mind (*unitas* and *mens*) as instances of individuals not composed of parts, Abbo goes further by explicitly asserting that God, like unity and the soul, is simple and lacks parts, being the individual which is the origin and the end of everything.

This conception of God as a simple entity without parts is in harmony with the Neopythagorean analogy between God and the arithmetical unity, as well as the unfolding of multiplicity as a 'flow'. To combine these three elements (i.e., God as *individuum*, as unity, and as a source from which multiplicity flows), Abbo uses a diagram similar to one developed by Calcidius to explain the psychogony, i.e., the generation of the World Soul in the *Timaeus* (Figure 4.1). In both Calcidius' and Abbo's diagrams, the shape resembles a

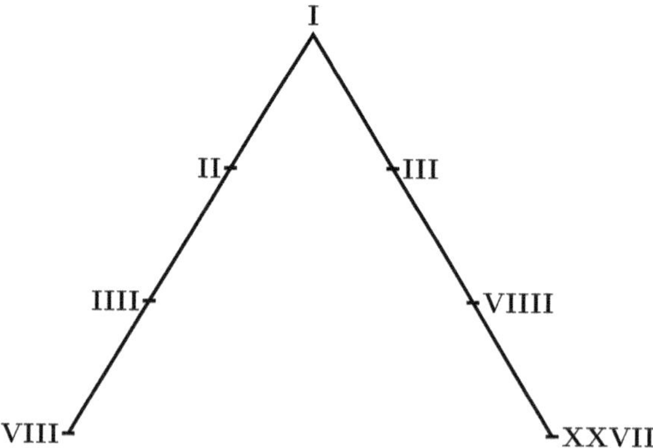

Figure 4.1 Re-elaboration of Calcidius' first lambda diagram from the manuscript Paris, BnF, lat. 2164, fol. 30v (end of 10th century, Fleury).

Greek lambda, with unity (*singularitas*) at the top, from which the first even (2) and odd numbers (3) descend to the left and right, respectively. These are followed by their respective squares (4 and 9) and cubes (8 and 27). Abbo openly draws an analogy between God and the unity at the top of the diagram, emphasising God as the principle of the multitude of creatures, much like unity is the principle of the multitude of numbers and ratios. Using Calcidius' words, Abbo describes the conception of unity related to this diagram:

> Indeed, a certain individual is the end and the beginning of all. Thus, Plato in the *Timaeus* found a suitable figure for the generation of the soul, where singularity is considered to occupy the summit and peak – like a pinnacle placed on the top – so that through it, a generous stream would flow out, like a sort of a river flowing from the bosom of the eternal source of the provident intelligence; and singularity itself can be understood as mind or intelligence or God. Since it is the origin of numbers, supplies substance to all things from itself, and contains their [*i.e.*, of the numbers] ratios (both simple and multiples), while other numbers change and depart from their own nature with increases and decreases, it alone, by a just and unshaken law, always remains the same in its state, always unchangeable and singular, just like all divine things that are not subject to the progress of time. What would exist firm if the foundation wavered? Geometers, in fact, call it a note, that is, a sign or point, which occupies its place undivided into any parts, and thus escapes the notice of the senses; nevertheless, it is perceived by the reason. From this note, a line is produced, as duality from unity, although the subsequent loses the power of the preceding, for neither is a line a sign, nor are two one. This is manifested by the aforesaid figure of psychogony, contained in the subtleties of arithmetic, geometry, music, and astronomy, and thus wonderfully expressed.[27]

The unity positioned at the top of the diagram, according to Abbo, can be understood as God. Abbo further characterises this unity as 'super-substantial' (*yperusian* or *supersubstantialis*), this terminology echoing Eriugena's characterisation of God as *superessentialis* – a transcendence beyond being itself, beyond any essential characterisations.[28] Besides this, it is evident that Abbo describes the origin of numbers from unity as a river flowing from a source. Thus, he clearly employs a vocabulary that aligns with the Neopythagorean tradition – and as we have seen, this tradition combines the idea of a flow of numbers from the arithmetical unity and the way created goods and substances derive their goodness and existence from a supreme good which is a simple principle. Abbo, thus, draws a parallel between unity and God. Similar to the divine principle, unity remains immutable even as numbers, ratios, and proportions 'flow' from it. This is indicated also by another statement by Abbo: "from the one that is the source of equality <ratio> [*i.e.*, 1:1] all the rivers of the inequality <ratios> flow gradually".[29] This claim hints at the idea that numbers forming an inequality ratio can be derived from those that share an equality ratio, a principle elucidated by Boethius in his *De arithmetica*.[30] The progression from a simple principle to what becomes more complex can also find expression through geometry, as the unity is associated with the point and what immediately proceeds from unity to the line – a parallelism that Abbo could find in both Calcidius' and Macrobius' works.[31] Starting from a point, a line can be derived, yet it loses its simplicity as it moves away from its origin. In this analogy, unity and the point correspond to God, the *individuum* from which everything originates. It is worth noting that arithmetic and geometry are not the only disciplines capable of expressing this analogy. The lambda diagram and its ratios not only have an arithmetical significance, but they potentially have an import in music and astronomy, as it is argued in the next section.

4.3 The Lambda Diagram: A 'Quadrivial' Henology

In addition to depicting arithmetic ratios, the lambda diagram introduced by Abbo also includes musical consonances, constituting a 'visual argument' that amalgamates insights from both arithmetic and music.[32] Through this diagram, Abbo refines his conceptualisation of God as *individuum*, signifying the origin and culmination of all things. Although the diagram is incorporated into the text and transcribed in all manuscripts of the *Explanatio*, the transcriptions vary among them. For the purpose of this discussion, I eschew reliance on Peden's critical edition of the *Explanatio* due to its lack of representation of harmonic ratios. Instead, it seems more appropriate to look at the diagram found in the manuscript Bamberg, Staatsbibliothek, Class. 53, fol. 11r (corresponding to Figure 4.2).[33]

To comprehend the significance of this diagram, let us first succinctly trace its original usage and development across the centuries. The diagram is intricately linked to Plato's *Timaeus*, a text familiar to medieval scholars through the partial translation by Calcidius.[34] In the Platonic dialogue, Timaeus of Locri, an alleged Pythagorean, narrates the demiurge's process of forming the World Soul. This involves blending the substances of Sameness, Difference, and Being, followed by structuring what results according to mathematical ratios through three principal steps.[35] Firstly, the demiurge initiated the division of the soul commencing with the initial and simplest part of it (1), then forming subsequent parts based on the double and triple geometric proportions (2^0, 2^1, 2^2, 2^3; and 3^0, 3^1, 3^2, 3^3). Thereby seven numerical values (1, 2, 3, 4, 8, 9, and 27) and six intervals between them (e.g., one interval is between 1 and 2, or between 2 and 3) were generated. The two series of numbers emanating from 1 through geometric proportion yield the first double and triple numbers (2 and 3), their squares (4 and 9), and their cubes (8 and 27).[36]

Secondly, the demiurge introduced two additional numerical values within the intervals of the double series (1, 2, 4, 8) and triple series (1, 3, 9, 27). For

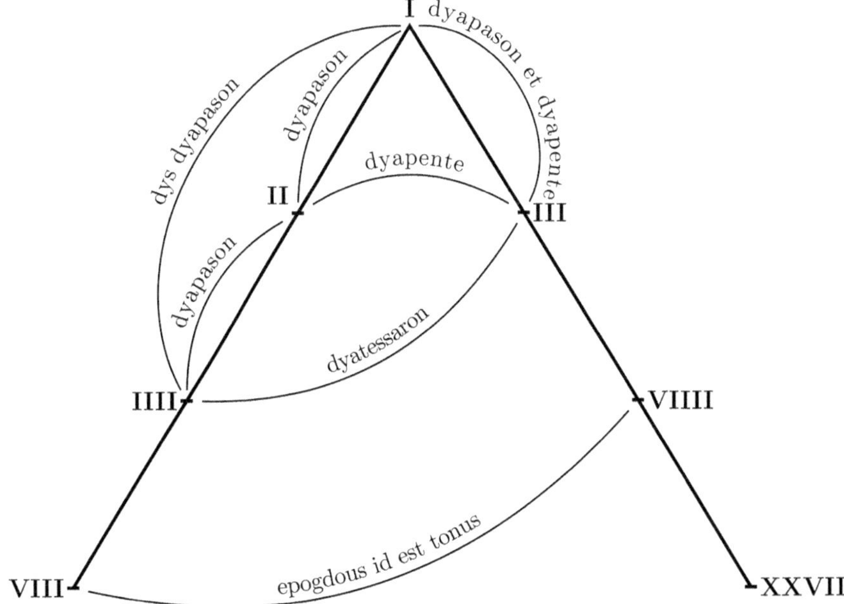

Figure 4.2 Abbo's lambda diagram. Re-elaboration based on Bamberg, Staatsbibliothek, Class. 53, fol. 11r.

each interval, two middle terms were identified based on the harmonic (2*ab*: *a* + *b*) and arithmetic proportions (*a* + *b*: 2). Consequently, in the double and triple series, the following values were produced, respectively:

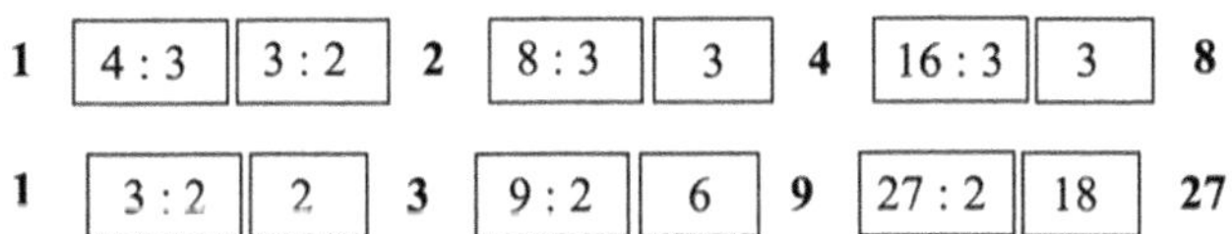

1	4 : 3	3 : 2	**2**	8 : 3	3	**4**	16 : 3	3	**8**
1	3 : 2	2	**3**	9 : 2	6	**9**	27 : 2	18	**27**

Finally, the demiurge filled all the intervals of 4:3 (i.e., the descending interval of the fourth according to Pythagorean music theory) with those of 9:8 (i.e., the tone). Consequently, each interval comprised two tones plus a tone fraction corresponding to 256:243 (referred to as *leimma*).[37] The proportions derived from this process constitute the intervals in the range of sounds of a diatonic scale. Yet one must acknowledge that, when Plato describes the generation of the World Soul, he does not imply any acoustic effects; for the harmony resulting from the demiurge's work was merely a noetic harmony – a purely mathematical harmony, one could say.[38]

The passage in Plato's work detailing the so-called *divisio animae* received special attention during the Imperial age, when the lambda diagram was developed to elucidate the arithmetical principles underlying the psychogony in the *Timaeus*.[39] In Plutarch's account, Crantor of Soli, the fourth Scholarch of the Academy (*fl.* the first half of the 3rd century BC), was the first to explain the generation of the World Soul using this 'visual argument'.[40] Later on, Adrastus of Aphrodisias, Theon of Smyrna, Plutarch, and Proclus also made reference to the diagram.[41] Calcidius can rightly be included in this list. In the prefatory letter of his *Commentary on Plato's Timaeus*, Calcidius expresses his intent to illuminate the technical aspects of the Platonic text.[42] Consequently, he emphasises the need to employ specific and technical explanations ("artificiosa ratio") drawn from disciplines such as arithmetic, astronomy, geometry, and music to clarify obscure issues present in the dialogue.[43] Mathematics, therefore, serves as one of Calcidius' primary tools for explaining and delving into Plato's *Timaeus*. This stands particularly for music. In Calcidius' view, the significance of both numbers and harmonies in the division of the World Soul by the demiurge is also underscored by Plato's choice of Timaeus of Locri as the one who expounds this doctrine. Having thoroughly studied Pythagoras' teachings, Timaeus was deemed an appropriate character to illustrate that there exists a precise correlation not only between the nature of the soul and numerical ratios but also between the soul and harmonies.[44] Since, for Calcidius, the nature of the soul is related to both numbers and harmonic proportions, the discipline of music becomes particularly useful for analysing the generation of the World Soul, which he exemplifies through three lambda diagrams.[45] The first of these diagrams, depicted below, forms the basis of Abbo's own diagram.

To elucidate the mathematical structure of the cosmic soul, Calcidius explores the interconnections between all the numbers that constitute the World Soul, namely, all the arithmetic ratios: the double (2:1), the triple (3:1), the sextuple or sesquialter (3:2), the quadruple (4:1), the epitriton or sesqui-third (4:3), and the sesqui-eight (9:8).[46] Then, he correlates these ratios with corresponding musical consonances, which, as we will see, are the same reported by Abbo in his diagram and can be summarised schematically as follows:

diapason (i.e., the octave, corresponding to the double ratio = 2:1);
diapason + *diapente* (i.e., the octave + fifth, corresponding to the triple ratio = 3:1);
diapente (i.e., the fifth, corresponding to the sesquialter ratio = 3:2);
disdiapason (i.e., two octaves, corresponding to the quadruple ratio = 4:1);
diatessaron (i.e., the fourth, corresponding to the *epitriton* or sesqui-third = 4:3);
tonus (i.e., the tone or *epogdoon*, corresponding to the sesqui-eight ratio = 9:1).[47]

In addition to Calcidius' *Commentary*, medieval scholars could find references to the arithmetic and harmonic proportions concerning the psychogony recounted in the *Timaeus* also in Macrobius' *Commentary on the Dream of Scipio*. Many of Macrobius' claims on the subject align with Calcidius'. Like Calcidius, Macrobius perceives a close connection between arithmetical and harmonic ratios, concluding that the World Soul's structure is in accordance with numbers that generate musical harmony.[48] Moreover, they both stress how numbers originate from unity, conceptualising this as a source from which the river of number flows – this suggests that Macrobius, too, could be placed within the Neopythagorean tradition as outlined in the preceding sections. In the arithmological excursus of the first book of his commentary, Macrobius delves into the properties of the numbers eight and seven. When these numbers are multiplied together, they yield the number of years associated with one of the characters in Cicero's narrative, Scipio Emilianus. The combination of eight and seven, according to Macrobius, signifies the union of even and odd numbers, which are fundamental features used to classify all numbers. Macrobius asserts that the significance of these features was acknowledged by Plato in the *Timaeus*, for the demiurge organises the parts of the World Soul in accordance with the principles of even and odd. From unity, the first even and odd numbers (2 and 3) are generated, along with their respective square (4 and 9) and cube (8 and 27) numbers.[49] Additionally, Macrobius suggests that the World Soul is formed based on numbers that provide solidity to things, with cubic numbers playing a significant role in representing this solidity. According to Macrobius, these numbers contribute to the animation of the entire universe, infusing life into everything it encompasses. The animation of a solid body is precisely conveyed through square and cubic numbers. More specifically, cubic numbers can be associated with a solid when considering geometric figures generated by the number of points bounding them: a line is defined by two points (2), a surface by four (2^2), and a solid by eight (2^3).[50] While Macrobius doesn't employ a diagram to elucidate the Timaean psychogony, he appears to allude to it through the following words:

> It was by this number first of all, indeed that the World Soul was begotten, as Plato's *Timaeus* has shown. With the monad located on the apex, two sets of three numbers each flowed on either side, on one the even, on the other the odd: that is, after the monad we had on one side two, four, and eight, and on the other three, nine, and twenty-seven; and the intertwining of these numbers brought about the generation of the World Soul at the behest of the creator.[51]

From this description, Macrobius indeed appears to be describing a lambda diagram.[52] According to the manuscript tradition, this diagram was not originally included in the text of Macrobius' commentary.[53] Yet, likely due to the philosophical affinity of this text with Timaean doctrines, medieval copyists transcribed the first of the three lambda diagrams elaborated by Calcidius (i.e., Figure 4.1) into Macrobius' text.[54] In two of the most ancient witnesses of the *Commentary on the Dream of Scipio*, namely, Paris, BnF, lat. 6370 and Paris, BnF, lat. 16677, our diagram is not included in the text but is still placed in the margin (either below the text or to one side).[55] This aspect remains unchanged in the manuscript tradition until the 11th century, albeit with some exceptions. One such exception is the manuscript Paris, BnF, lat. 6365, which was copied at Fleury and dates back to Abbo's time.[56] One might therefore ask whether Abbo played a role in the inclusion of the diagram in the text of Macrobius' commentary. Manuscript evidence corroborates the possibility that he took into account not only Calcidius' lambda diagram but also Macrobius' considerations about the Timaean psychogony, making the idea that he contributed to the inclusion of the diagram in the main text of Macrobius' work very likely. Let us consider manuscript Paris, BnF, lat. 2164 and the manuscript Paris, BnF, lat. 6365. As shown by Michel Huglo, these are 'twins' manuscripts, being copied at the same time and in the same scriptorium, at Fleury.[57] Their contents were originally arranged in the following order: Mamertus' *De statu animae*, Calcidius' translation and commentary on the *Timaeus*, and Macrobius' *Commentary on the Dream of Scipio*.[58] The fact that in his *Explanatio*, Abbo mentions both Calcidius and Macrobius with respect to his own lambda diagram aligns with the codicological evidence that the two above-mentioned manuscripts come from Fleury.[59] In both Calcidius' and Macrobius' works transmitted in these manuscripts, one can find the lambda diagram. Surprisingly, the diagram is incorporated within the text and not put in the margin of Macrobius' *Commentary*. Although I cannot definitively answer the question of whether Abbo played a role in the inclusion of the lambda diagram in the text of Macrobius' *Commentary*, one fact remains. Abbo studied and reflected on Calcidius' and Macrobius' passages dealing with the Timaean psychogony, and did so in such an original way that he elaborated his own diagram (Figure 4.2) and mediated between the two philosophical authorities in building his own henology.

The shape of Abbo's diagram is a Greek lambda, incorporating the same numerical values already described (i.e., 1, 2, 3, 4, 8, 9, 27). However, the remarkable originality of Abbo's diagram lies in a distinctive feature: musical consonances are also included, for the numbers are linked by arcs of circles. In line with Boethius' *De institutione musica*, Abbo designates the ratio 2:1 as *dyapason*, the 4:1 as *bis dyapason*, the 3:2 as *dyapente*, the 3:1 as *dyapason plus dyapente*, the 4:3 as *dyatessaron*, and the 9:8 as *tonus*. This nomenclature establishes a link to the character of Pythagoras and his role as the discoverer of the discipline of music. Similar to Nicomachus in his *Manual of Harmonics*, both Boethius and Macrobius attribute the discovery of the consonances to Pythagoras and recount the same narrative. According to the story, Pythagoras, while passing by a blacksmith's shop, discerned a harmony in the sounds produced by the hammers striking the iron. Upon closer observation, Pythagoras connected the different sounds to the varying weights of the hammers. He noted that hammers weighing twice as much as others produced a *diapason* consonance, those weighing sesqui-third produced a *diatessaron* consonance, and those weighing sesquialter produced a *diapente* consonance.[60]

In contrast to the manuscript from Bamberg (on which Figure 4.2 is based), in some other manuscripts of the *Explanatio*, the arches connecting the numbers are drawn, but only arithmetical ratios are mentioned along them. Nevertheless, all lambda diagrams in the extant manuscripts of Abbo's *Explanatio* are accompanied by brief explanations that allow for converting arithmetical ratios into musical consonances: "Duplus est diapason. Triplus est diapason et diapente. Bis duplus bis diapason. Epogdous id est sesquioctavus id est tonus".[61] These musical remarks complementing the diagram would confirm Michel Huglo's opinion that the *Explanatio* also aimed to clarify the subdivisions of the unity discussed in Boethius' *De institutione musica*.[62] Since arches do not appear in the diagrams of Calcidius and Macrobius' works, to comprehend Abbo's diagram fully, it is indeed crucial to refer to Boethius' musical textbook, particularly his diagrams, in which the space of musical intervals is precisely expressed through arches connecting the numbers constituting an arithmetical ratio. The harmonic ratios presented by Abbo in the diagram express intervals, representing the difference in pitch between lower and higher sounds. The mathematical nature of these intervals was thoroughly examined by Boethius in his *De institutione musica*. According to Boethius, the focus of the music discipline is on consonance, which refers to the harmony of dissimilar sounds brought back to unity.[63] The ratios that express consonance and contribute to the reduction of sounds to unity ("reductio in unum") include the simple (multiple) ratios, such as double, triple, and quadruple, as well as the superparticular ratios like sesquialter and sesqui-third.[64] As evident from Figure 4.2, Abbo's diagram incorporates all these five ratios outlined by Boethius. For Boethius, the comprehensibility of sounds relies on their adherence to a specific ratio (a: b) that defines the interval. Proportions (a: b = c: d) result from the combination of intervals and can be deconstructed into the individual ratios that constitute them (a: b; and c: d). In turn, ratios can be further simplified into more basic ratios. The ultimate goal of this process is to reduce every ratio to the equality ratio (1:1), representing unison, which symbolises musical perfection. The closer the terms of a proportion are to the equality ratio, the higher the perfection of that proportion. In this context, the diapason consonance, characterised by the ratio 2:1 (equivalent to an octave), stands out as the nearest to this perfect ratio.[65]

In light of these observations on Boethius' music theory, it is noteworthy that Abbo's diagram places the equality ratio at the pinnacle, serving as the point towards which all consonances tend and to which they are ultimately reduced. Emphasising the significance of music in Abbo's 'visual argument', it is essential to recognise that, unlike Calcidius and Macrobius, the diagram in the *Explanatio* is not introduced to elucidate the Timaean psychogony. Instead, it illustrates how arithmetical and harmonic ratios can both flow from and be traced back to unity and the equality ratio. Unison, containing diverse consonances, becomes a key element in Abbo's argument, thus expressing his henology viewed through the lens of musical discipline. God, as *individuum* from and in whom everything originates and return, becomes analogous to the unison in this context. Therefore, one could see Abbo's henology as a sort of updated or expanded version of henology, for it is grounded not only on arithmetical principles but also on musical ones.

Abbo openly expresses the contribution of music in understanding the diagram, stating that it can be interpreted through all mathematical disciplines traditionally constituting the *quadrivium*, including arithmetic, geometry, music, and astronomy.[66] Taking this statement seriously involves identifying a principle from each mathematical discipline that can be associated with the divine one. As it has been shown, in the context of arithmetic and music, suitable principles are readily available. Moreover, in geometry, too, the entity of the point can certainly play a role, as Abbo himself suggests when assimilating unity and point – with the line originating from the point somehow reflecting the numbers and ratios originating from unity and unison.[67] However, the role of astronomy is less clear in the *Explanatio*. Abbo may have mentioned it simply to complete the list of mathematical disciplines, or he might have referred to a specific astronomical issue that contributes to elucidating his diagram and henology.[68] To explore this further, it is necessary to examine what Abbo's sources say about his 'working table'.

Convinced of the harmony between the mechanics of the heavens and acoustic principles, both Calcidius and Macrobius explored the proportional spatial arrangement of the planets based on musical intervals. According to Macrobius, the World Soul imparts a circular motion to the world's body, creating a sound grounded in inaudible intervals graspable only by reason. These intervals are said to be preserved in the structure of the world, specifically in the arrangement of the planets.[69] Macrobius identifies an initial spatial distance (Earth-Moon) and argues that each subsequent planetary distance must be a multiple of the previous one. This results in the same continuous geometric proportion as the numbers structuring the World Soul (1, 2, 3, 4, 8, 9, 27). The space between the Earth and the Moon corresponds to one, the space between the Earth and the Sun is twice the Earth-Moon distance, the space between the Earth and Mercury is three times the Earth-Sun distance, and so on, up to the distance between the Earth and Saturn, which is 27 times the Earth-Jupiter distance.[70] Calcidius, on the other hand, presents a similar sequence of planets but conceives each space between them as a multiple of the initial Earth-Moon distance.[71] The theories discussed by Calcidius and Macrobius, which were most likely known to Abbo, are also explored in a musical treatise titled *Dulce ingenium musicae*, attributed to Abbo by Michel Huglo. In this text, Macrobius and Boethius are referenced regarding the planetary arrangement and the definition of *musica naturalis*: "The nerves, that are the strings of artificial music, can be compared to celestial music. In fact, the *hypate meson* [*scil.* the lowest sound] is attributed to Saturn and the other <sounds> to the seven planets, although with a different order according to Boethius and Macrobius".[72] In light of this reference to Macrobius, it can be conjectured that Abbo had in mind an astronomical diagram similar to the one found in the Fleurisian manuscript Paris, BnF, lat. 6370, preserving the *Commentary on the Dream of Scipio* with glosses by Lupus of Ferrières and Heiric of Auxerre. On folio 76r, a diagram of heavenly bodies (Figure 4.3) clarifies Macrobius' statements on planetary intervals (*In Somn.* II, 3, 11–14).

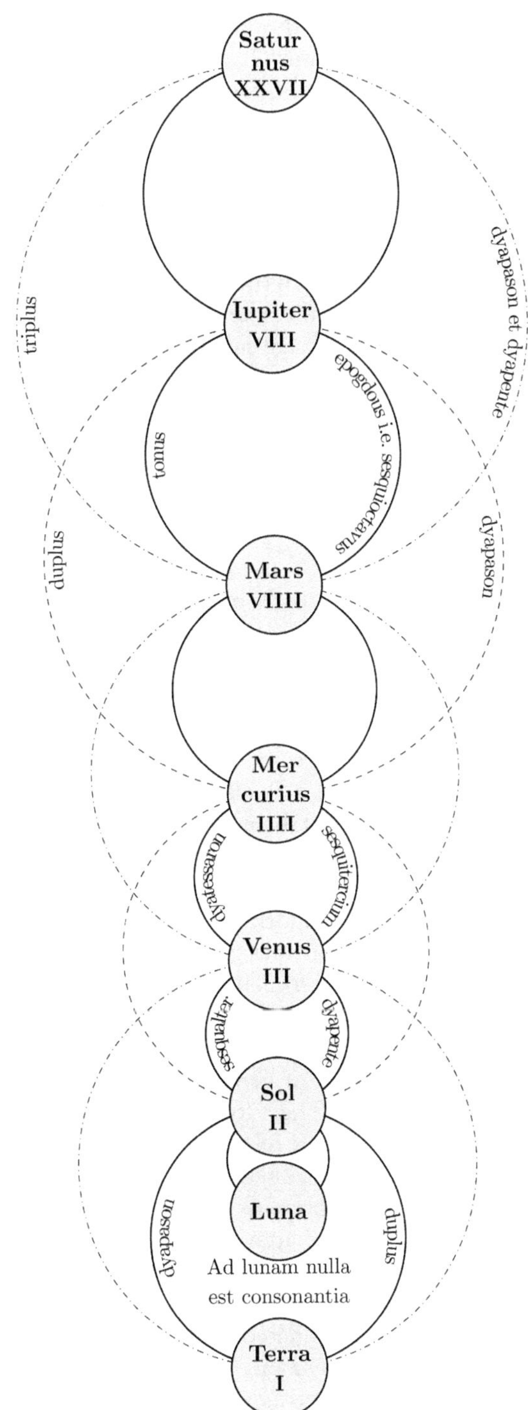

Figure 4.3 Re-elaboration of the diagram from Paris, BnF, lat. 6370, fol. 76r.

To summarise, Abbo employs the lambda diagram as a 'visual argument' in which he considers the unit (1) and the equality ratio (1:1) as the sources of numbers and consonances, respectively. This approach is rooted in the insights of Calcidius and Macrobius on the Timaean psychogony. Additionally, a potential explanation for the reference to the astronomical discipline in connection with the diagram is to be found in astronomical diagrams such as those in manuscript Paris, BnF, lat. 6370, where arches connect planets based on musical and arithmetic ratios.[73] However, while the astronomical diagram aligns with the mathematical rationale of lambda diagrams, it fails to fully incorporate a henological perspective. In contrast, the metaphysical and theological significance of unity is readily apparent in the arithmetical and musical diagrams – and even within a geometric context, it is evident in the reference to the point-line dynamic. This leaves us questioning why such significance should be associated with the Earth in the astronomical context.

4.4 Unity as Ontological Principle

Abbo's understanding of unity extends beyond the analogy with God and the conceptualisation via mathematical disciplines, for it also encompasses the role of unity as the principle of (individual) existence. For unity is also seen as ensuring everything exists as a singular entity. This conception is to be placed against the logical backdrop influenced by Boethius, where the terms 'being' and 'being one' are treated as logically interchangeable. In his *Second Commentary on the Isagoge*, Boethius supports Porphyry's assertion that 'being' (*ens*) cannot function as the most general genus encompassing all ten categories. The predication of 'being' across different categories relies on equivocation, signifying that it is not used in a univocal sense when applied to terms falling under diverse categories.[74] Boethius further contends that 'being' cannot be conceived as a distinct genus above the categories because 'being' is equivalent to 'one' (*unum*). This equivalence leads to the logical interchangeability of 'being' and 'one' when predicated of things. However, Boethius asserts that they cannot be considered genera with respect to the categories, nor can they be species or genus of each other – 'one' is not the genus of 'being', and vice versa. Their logical interchangeability sets them apart from traditional genus-species relationships.[75] Furthermore, Boethius emphasises the correlation between 'being' and 'being one' in the context of the problem of universals. He presents an aporia that challenges the real existence of genus and species. The aporia arises from the presence of each universal in many things simultaneously, preventing it from having unity, with the consequence that it cannot be a singular entity in number. Since only that which is one in number is considered to exist, the universal, lacking this singularity, cannot be conceived as having real existence. Boethius introduces the principle that everything that exists is also one precisely to address this aporia.[76]

The equivalence of 'being' and 'being one' is highlighted also in Boethius' *Contra Euthychen et Nestorium*. This theological treatise addresses the heresies of Eutyches and Nestorius, who held different views on the nature of Christ. While both juxtaposing the concepts of 'nature' and 'person', Eutyches argued for a single divine nature in Christ, resulting in only one person, while Nestorius claimed a dual nature (divine and human) and, consequently, two persons in Christ. Boethius counters Nestorius' position by asserting the mutual conversion of 'being' (*esse*) and 'one' (*unum*). He argues that unity is essential for the existence of anything and thus, there must be only one person in Christ: if there were two persons in Christ, the unity of his being would be lacking, and Christ would not exist at all.[77]

The Boethian parallel of 'being' and 'being one' continued to have a significant impact throughout the Middle Ages. With the renewed interest in some of Boethius' theological opuscules starting from the 12th century, during the 13th century, scholars began to incorporate the Boethian parallel into a more systematic theory of 'transcendentals'.[78] However, even in the 9th century, there is evidence of an embryonic and concise reflection on Boethius' equivalence of 'being' and 'being one'. In a set of anonymous glosses to the *Contra Eutychen*, attributed to John Scotus Eriugena by Edward Rand, the commentator focuses on the passage regarding the convertibility of *esse* and *unum*.[79] The author paraphrases Boethius' words, stating: "Reciprocatur ita: quodcunque est, unum est, et quodcunque unum est, illud est" – "It is reciprocated in this way: whatever is, is one, and whatever is one, is".[80] Although this reflection is extremely concise, it nevertheless indicates an early interest in – or at least an awareness of – the logical principle of conversion of 'being' and 'being one' in the 9th century.

At the turn of the first Millennium, particularly within Abbo's *Explanatio*, the Boethian theory of the logical conversion between 'being' (*ens/esse*) and 'one' (*unum*) seems to gain more prominence. Abbo incorporates this theory into his henology, also developing a broader discussion on unity as an ontological principle. According to Abbo, those who investigate the nature of unity must first acknowledge it as a worldly cause ("ratio mundani principii"). With this expression, Abbo refers to a cause or principle that encompasses all of created reality, and through which everything that exists does so insofar as it is one in number. More specifically, Abbo argues that 'is' (*est*) and 'one' (*unum*) are convertible when predicated, so that 'is' and 'one' have same semantic extension. This implies that it is not possible to say that something that is not is one, nor that something that is not one is. Furthermore, Abbo engages with the same aporia concerning universals presented by Boethius in his *Second Commentary on the Isagoge*. Abbo's reasoning begins with the premise that everything that truly exists must be one in number. He acknowledges that universals, although said to be present in multiple things simultaneously, are nevertheless said 'to be' in a certain sense, but not in the same way as the individual concrete entities. For example, the universal 'human being' does not exist in the same way as Socrates does, precisely because only

Socrates, that is, the individual human being is truly one in number ("causa unitatis").81

> Those who investigate the nature of unity put forward for consideration some worldly cause. And it must be said first why unity does not admit of division. Everything that is, is one; and whatever is one, it is necessary that it be. Indeed, 'one' and 'is' are mutually convertible in their predication. Those things which are convertible in their predication are equal. Therefore, 'one' and 'is' are equal, because neither 'one' can be said it is not, nor 'is' that it is not one – although there are certain universal things which, while being wholly in several things at the same time, are none the less said to exist, like 'human being' who is wholly in Socrates and Plato at the same time, from which it gathers its existence. Therefore, it follows that 'human being' understood universally is not, as it always exists as particular [individuals] on account of unity.[82]

In Abbo's perspective, asserting the existence of something is equivalent to asserting its oneness, and that it exists precisely because it is one. The emphasis on the concrete existing individual is also highlighted by the interchangeability of the third-person singular verb to be 'is' (*est*) and the term 'one' (*unum*). This former term of the equation notably differs from that employed by Boethius. In the *Second Commentary on the Isagoge*, Boethius introduced the conversion of 'being' (*ens*) and 'one' (*unum*), while in the *Contra Euthychen et Nestorium*, he likened 'to be' (*esse*) and 'one' (*unum*). This discrepancy could be understood by considering the distinct nature of these two works by Boethius – one being logical and the other theological. In the logical work, *ens* is mentioned concerning the status of genera and species, while in the theological work, *esse* is used in reference to the person of Christ.[83] On his part, Abbo draws a comparison between *est* and *unum* and deliberately opts for the singular person of the verb 'to be', emphasising the subject of such a verb, which is the thing grasped in its concrete particularity and singularity.

Abbo is aware that being one does not necessarily rule out the possibility for something to be concretely cut up into smaller parts. Rather, he associates the impossibility of being divided to the word 'unity' itself and, more broadly, to the names we use for counting or numbering things. Division does not affect words expressing enumerability. For instance, when we consider the term 'duality', which might intuitively suggest a division by two, the word itself remains undivided and implies no concrete division. On the contrary, concrete things that are numbered and named by such words can be physically divided. Yet, in the process of division, these entities lose their integrity and cease to be considered as wholes, as being one. For instance, when a part is added or subtracted from an entity measuring one foot, that entity is inevitably altered; yet, the 'unity' by which we counted it as one is not altered.

> It should be understood in counting that there is a difference, for the thing by which one counts is different from the thing which is counted. The thing by which one counts is the name of the numbers, like 'unity'; but that which is counted is denominated by a term, like 'one' from 'unity'. The same applies to 'two' and 'duality', 'three' and 'trinity', and so on. The name of a thing never undergoes increase or decrease, even though things denominated come in a comparison of degrees. They [*i.e.*, the names], in a way, follow the basic rule of accidents, which, just as they do not subsist by themselves, so they cannot suffer anything by themselves. Therefore, if 'duality' avoids division by itself, how much more so does 'unity'? But the reason does not regard the unity of a substance as something to divide, but it does so with duality, yet it cannot [divide it] by itself. Therefore, in the realm of things that are countable, the division of two is straightforward as is the division at any point of one, but in such a way that, for example, if there is some thing measuring one foot or something else similar in measure or weight, nothing can be added or subtracted from it by dividing it, except by altering the quality of the former unity. We do not assert that nothing can be added to or subtracted from [something measuring] a pound or a foot, but that that which is one does not require addition or subtraction to remain a single entity.[84]

Things that are numbered and named by the word *unitas* can be certainly partitioned, and yet not without varying their concrete unity, namely, their being a whole or their being an individual substance. In other words, when something *unum* is considered as a concrete continuous magnitude, that is as a whole, it can increase or decrease in quantity. However, such a change comes at the cost of losing its intrinsic singularity and ceasing to exist as *unum*. In other words, its unity is compromised when it is divided or altered, and so is its being, its existence as singular.

4.5 Unity as the Object of Arithmetic and Calculus

Further components shaping Abbo's henology as referred to created things are number theory and calculus. In this perspective, number is viewed as a collection of units and unity as the fundamental principle underlying the plurality of numbers. To address this matter, reference must be made to Boethius' *De arithmetica* and his definitions of multitude and magnitude, which serve as the basis for Abbo's reflections.[85] The notion of number is encompassed within the definition of multitude, representing every discrete quantity composed of individual, separate, and independent parts. Examples are a flock, a group of people, and number itself – these being all a sum of singular units

(1 + 1 + 1 + ...). Conversely, magnitude pertains to continuous quantities whose parts are not distinctly identifiable (given their contiguity and continuity) and independent from one another. Examples include a stone, a person, or any other physical entity. While multitude has the potential for indefinite expansion, magnitude can be infinitely divided. These Boethian definitions are expounded upon in the *Explanatio*, in order to make sense of Victorius' claims that unity is that form which the multitude of numbers proceeds.[86]

> But since among all things, some are concrete and compacted (like a human or a stone), some others are dispersed and determined by their parts (like a flock or a crowd), for that reason Victorius adds "from which all the multitude of numbers proceeds" after "unity". [...] Multitude pertains to discrete and disaggregated things, just like magnitude pertains to continuous and compacted things, as multitude grows from one to infinity and magnitude decreases infinitely. The term 'multitude' always signifies the collection of plurality, and by the term 'magnitude', the quantity of each body is designated [...]. Since number is the collection of units and, in a way, the heaping of many parts into one, it is rightly said of it that it is a multitude.[87]

In the *De arithmetica*, Boethius distinguished two ways in which a multitude can be viewed: depending on whether quantity is considered in itself (*per se*) or in relation to something (*ad aliud*). Multiplicity considered in itself is the subject matter of arithmetic, while multiplicity related to other quantities is investigated by music. Both arithmetic and music explore number, with the former examining unrelated numbers (e.g., one, two, three, four, etc.), and the latter exploring numbers in relation to others (e.g., double, half, triple, etc.). The same distinction applies to magnitude. It can be considered either at rest or in motion. Hence, it becomes the object of study in geometry and astronomy, respectively.[88]

Unlike Boethius, Abbo posits that the proper focus of arithmetic is not the multitude itself but rather the unity that gives rise to multiplicity. This interpretation is rooted in his understanding of the opening sentence of Victorius' *Praefatio* to the *Calculus*: "the unity from which the entire multitude of numbers proceeds, and which properly belongs to the discipline of arithmetic, indeed admits no division".[89] While Victorius likely refers the clause "which properly belongs to the discipline of arithmetic" to the multitude itself, Abbo links the relative pronoun 'which' (*quae*) to unity, making it the object of arithmetic. According to Abbo, unity serves two essential functions in relation to other numbers: (1) it generates them and (2) it serves as their 'common measure'. In the arithmetic theory of the time, acting as a common measure means being a factor shared by two or more numbers. Unity is the factor of every number, whether commensurable or not. As illustrated in the following passage, Abbo defines 'measure' as the factor that, when multiplied *n*-times, produces a number. From this perspective, every number is a multiple of the initial unity, which in turn is the proper object of arithmetic.

> Unity pertains to geometry, music, and astronomy, but specifically to arithmetic, as number derives from it and it is called the measure of numbers. A number is measured either by unity alone, which is the common measure of all, as in the case of three and five; or by some number multiplied by its quantity, as four [is measured] by two, and nine by three; or not only by one but also by two numbers, as sixteen, which is measured by four, two, and eight. [...] Some numbers are composed of multiple [measures], such as twelve, which is brought to completion by one, six (as its second part), two (as its sixth part), and three (as its fourth part). [...] There are also commensurable and incommensurable numbers. Commensurable numbers are those that share a common measure, such as nine, twelve, and fifteen. [...] Incommensurable numbers are those that, except for unity, do not share any common measure, such as eight and nine.[90]

Unity can also be regarded as the 'first part' of each number, with 'part' signifying a fraction or aliquot. In the case of the number 12, 1 is its first aliquot part ($\frac{1}{12}$), 6 is its second (equivalent to $\frac{1}{2}$), 2 is its sixth ($\frac{1}{6}$), 4 is its third ($\frac{1}{3}$), and 3 is its fourth ($\frac{1}{4}$).[91] In contemporary mathematical language, expressing unity as the first part of every number involves placing the number one as the numerator of a fraction. For instance, the first part of 48 is the first unity that constitutes this number ($\frac{1}{48}$); the first part of 72 is $\frac{1}{72}$, and so on for any other number. This conceptualisation of unity as the first aliquot pertains to calculus. This stands also when measuring or dividing a whole, a continuous quantity, such as a line, whose length is considered an integer number.

> Often indeed, any part of a line or of any other thing is sought by those less skilled [...] and sometimes they fail in their frustrated labor while examining it unit by unit. In order for it not to happen, first, the parts of the number by which you attempt to divide the body must be investigated. And it should be understood that as many divisions as that number undertakes, so many are received by the body subject to that number. Like in the case of the twelfth part of a line: taking, if necessary, a compass, and placing one end of it at the end of the line and the other at the midpoint, you will find six units on one side and six on the other. You will likewise examine the half of these six units using the compass. Upon finding three [units], which number cannot be further divided because it is an odd number, you use the compass again to distribute the three into single units. Therefore, the third part of this three [units], in relation to the total length of the measured body, represents one twelfth of the twelve-unit whole.[92]

Abbo's explanation of the division of a line can be illustrated through the following example. Consider a line AB, where AB = 12, and one is seeking its 'first part' – the unit corresponding to $\frac{1}{12}$ of the entire line. According to Abbo's procedure, the line division unfolds as follows: using a compass, the segment AB is bisected precisely at its midpoint, creating point C, with AC = 6 and CB = 6. Next, these two segments are each halved at points D and E, resulting in AD = DC = CE = EB = 3. Although it is not feasible to divide this last portion into two equal parts represented by two integer numbers, one can use the compass to identify three equal parts of this last section. The outcome of these divisions yields the 12th part of AB, which is the 'first part' corresponding to unity.[93]

Abbo presents an alternative method for dissecting a body and reaching this elementary part, which is unity. Even a segment whose size corresponds to an uneven number can be divided in half using a different process that does not involve the use of a compass. Instead, it considers only the number corresponding to the size of a body, progressing incrementally or *unciatim*, literally 'ounce by ounce'. For example, starting from a body or segment corresponding to 17, its half (i.e., 8 and $\frac{1}{2}$) can be obtained. Then, the half of 8 and $\frac{1}{2}$ is 4 and $\frac{1}{4}$, followed by the half of 4 and $\frac{1}{4}$, which is 2 and $\frac{1}{8}$, and so on, until reaching $\frac{1}{16}$, which corresponds to $1 + \frac{1}{24} + \frac{1}{48}$. However, through this *unciatim* division of odd numbers and their corresponding bodies or segments, it is not possible to obtain the 'first part', namely unity. Instead, one can only approach unity while still retaining a fraction of the remainder (which, in the case of 17, is $\frac{1}{16}$).

> The halving of an odd [number] is done in the following way: the half of seventeen is 8 and a half. The half of this half is again 4 and a quarter; the half of this is 2 and a sixth, and the half of this is 1 with a twelfth and a twenty-fourth. Thus, unity of an odd number that disrupts its equal division is gradually reduced to equality in the body, fraction by fraction; however, it never results in a simple division through halving alone.[94]

Abbo emphasises, through these examples, the correlation between the division of numbers and the partitioning of a body. In this context, a smaller denominator, like the 2 in the case of $\frac{1}{2}$, clearly signifies a larger quantity in bodies. Conversely, a larger denominator, such as 8 in the case of $\frac{1}{8}$, corresponds to a smaller quantity. Abbo concludes by stating that even children understand that receiving one-third of a cake (i.e., $\frac{1}{3}$) is more desirable than its eighth part (i.e., $\frac{1}{8}$).

> When you divide or compare numbers, do not think you are doing anything to the body itself, unless you are using the numbers merely as a tool to measure some span of time. Indeed, whether a whole body is divided without cutting or with cutting, it is clear – even to children – that a smaller number represents larger parts of the body, and a larger number represents smaller parts; hungry children prefer a third of a

loaf of bread to a fourth, and a fourth to an eighth. Indeed, whenever you have divided something by cutting, undoubtedly the plurality of numbers increases in the manner of multiplicity, while the quantity of the body decreases in the manner of magnitude.[95]

A further message that Abbo seems to convey is that the division of numbers finds correspondence in the breakdown of physical wholes. This belief is articulated also in one of his astronomical works, specifically the *De cursu septem planetarum*, where it is suggested that division by calculus has tangible implications in the realm of bodies: "Et idcirco dum numeros dividis, multiplicas, non nihil agere te credas".[96] Such is the focus of the next chapter.

Notes

1 I follow the points delineated in Albertson, *Mathematical Theologies*, 23–27.
2 The following analysis is not extensive, for I deemed it prudent to focus only on those thinkers who thought of the divine principle in explicit juxtaposition with the arithmetic unit. To address any gap left open, one can refer to Bonazzi, "Plotino", as well as to the collective volume Bonazzi, Lévy, and Steel, *A Platonic Pythagoras*. For a comprehensive understanding of the various ways in which the Pythagorean tradition permeates the Middle Ages, it is essential to consider Hicks, "Pythagoras", and Caiazzo, Macris, and Robert, *Brill's Companion*.
3 Nicomachus of Gerasa, *Introductionis arithmeticae*, I, 7, 1, p. 13; translation in Nicomachus of Gerasa, *Introduction to Arithmetic*, 190: "Number is limited multitude or a combination of units or a flow of quantity made up of units". According to Nicomachus, the arithmetical unity, possessing the potentiality of encompassing all numbers, is not merely a number in itself. Analogously, he likened the point to the arithmetical unity: the point is the possibility of all lines, to not being a line in itself. See Nicomachus of Gerasa, *Introductionis arithmeticae*, II, 6, 3; translation in Nicomachus, *Introduction to Arithmetic*, 237: "Unity, then, occupying the place and character of a point, will be the beginning of intervals and of number, but not itself an interval or a number, just as the point is the beginning of a line, or an interval, but is not itself line or interval". For the following exposition of the Nicomachean reflection, I refer to the analysis by Robbins and Karpinski, "Studies in Greek Arithmetic".
4 Nicomachus of Gerasa *apud* Iamblichus, *Theologoumena arithmeticae*, 1, p. 3; translation in Iamblichus, *The Theology of Arithmetic*, 37: "Furthermore, the monad produces itself and is produced from itself, since it is self-sufficient and has no power set over it and is everlasting; and it is evidently the cause of permanence, just as God is thought to be in the case of actual physical things, and to be the preserver and maintainer of natures". On Nicomachus' parallelism, see also M. L. D'ooge, *Studies in Greek Arithmetic*, in Nicomachus, *Introduction to Arithmetic*, 96.
5 *Arithm.*, I, 7, 5–6, p. 16: "Quare constat prima esse unitatem cunctorum qui sunt in naturale dispositione numerorum et eam rite totius quamvis prolixae genitricem pluralitatis agnosci"; *Arithm.* I, 2, 4, p. 12: "Haec autem sunt quibus numerus constat: par atque impar, quae divina quadam potentia, cum disparia sint contrariaque, ex una tamen genitura profluunt et in unam compositionem modulationemque iunguntur"; *Arithm.*, I, 3, 2, p. 12: "Numerus est unitatum collectio, vel quantitatis acervus ex unitatibus profusus". See also Guillaumin, "Boèce traducteur".

6 *Arithm.*, I, 14, 2, p. 31: "Dicitur autem primus et incompositus quod nullus eum alter numerus metiatur praeter solam quae cunctis mater est unitatem". As we will see, this is one of the reasons why Abbo believes that unity is the specific object of the arithmetic discipline.
7 Boethius, *De institutione musica*, II, 5, p. 246: "Geometrica vero proportionalitas tunc quemadmodum inveniri ab aequalitate possit ostendimus, quando, quemadmodum ab aequalitate omnis inaequalitas profluat, monstrabamus". Hereafter *Musica*. For a more detailed explanation of this arithmetical procedure see Chapter 5 in this book.
8 Albertson's view (*Mathematical Theologies*, 86) is different: "And yet viewed as a totality, Boethius's collection of Neopythagorean fragments never interacts in a fundamental way with his high views of the Trinity and the Incarnation".
9 Other (Neo)Pythagorean elements can be found in the *De hebdomadibus*. For instance, according to Dominic O'Meara, the axiomatic method of the *De hebdomadibus* provided medieval theologians with a 'Pythagorean' model of scientific discourse; O'Meara, *Pythagoras Revived*, 210–211: "One might mention as an example [of Pythagoreanizing program] Boethius' short theological work known as the *De hebdomadibus* which would provide, in its geometrical form, a model for scientific discourse to Western Medieval theologians". See also Galonnier, "Axiomatique"; and Micaelli, "Boethius's *De Hebdomadibus*". Additionally, the title itself may contain an esoteric reference to Neopythagorean theories of numbers and their symbolic value, especially the number seven. This is the thesis of Pessin, "Hebdomads". Boethius' reference to the *hebdomad*, namely to seven, might rest precisely on Nicomachus' idea that seven represents the divine 'limb' or God's main instrument in the creation. See Nicomachus, *Theologoumena arithmeticae*, 7, pp. 57–58; translation in Iamblichus, *The Theology of Arithmetic*, 89–90. According to Jean-Luc Solère (Solère, "Bien"), Proclus would be at the origin of Boethius' use of the term *hebdomas*. It is worth reminding, however, that 'hebdomads' was the title of the work by John, a Deacon who happened to be the dedicatee of Boethius' work. Among the issues John dealt with, there is the same Boethius grapples with in his *De hebdomadibus*: whether created substances are good insofar as they are.
10 For a comprehensive survey of Boethius's arguments, I refer to McDonald, "Boethius's Claim"; and to Hadot, "La distinction".
11 *Hebd.*, 188, ll. 41–42: "Omne simplex esse suum et id quod est unum habet". Again on the identity of being and acting, see *Hebd.*, 193, ll. 156–159: "Idem autem est in eo esse quod agere; idem igitur bonum esse quod iustum. Nobis vero non est idem esse quod agere; non enim simplices sumus". In the *Consolatio*, too, Boethius expressly refers to God as *fons bonorum*; cf. Boethius, *Consolatio*, III, m. 9, 23, p. 80: "Da, pater, augustam menti conscendere sedem, // da fontem lustrare boni". Hereafter *Consolatio*. See also *Consolatio*, III, pr. 10, 3, p. 81: "Sed quin existat sitque hoc veluti quidam omnium fons bonorum, negari nequit". Again in Book III of the *Consolatio*, Philosophy explains to Boethius that all particular goods are such in that they participate in a single principle which is the good in itself, envisaging a parallel not only between what is absolutely one, simple and good by nature, but also true; *Consolatio*, III, pr. 9, 4, p. 76: "Quod enim simplex est individumque natura, id error humanus separat et a vero atque perfecto ad falsum imperfectumque traducit". For these issues, see Zambon, "La ricerca", 150–155.
12 *Hebd.*, 191, ll. 108–117: "Quae [i.e. substantiae] quoniam non sunt simplicia, nec esse omnino poterant, nisi ea id quod solum bonum est esse voluisset, idcirco, quoniam esse eorum a boni voluntate *defluxit*, bona esse dicuntur. Primum enim bonum, quoniam est, in eo quod est bonum est; secundum vero bonum, quoniam ex eo *fluxit* cuius ipsum esse bonum est, ipsum quoque bonum est. Sed ipsum esse

omnium rerum ex eo *fluxit* quod est primum bonum et quod bonum tale est ut recte dicatur, in eo quod est, esse bonum". My italics.
13 *Trin.*, II–III, 170, ll. 92–124: "Sed divina substantia sine materia forma est atque ideo unum est, et est id quod est: reliqua enim non sunt id quod sunt. Unumquodque enim habet esse suum ex his ex quibus est, id est ex partibus suis, et est hoc atque hoc, id est partes suae coniunctae, sed non hoc vel hoc singulariter, ut cum homo terrenus constet ex anima corporeque, corpus et anima est, non vel corpus vel anima in partem; [...]. Quod vero non est ex hoc atque hoc, sed tantum est hoc, illud vere est id quod est; et est pulcherrimum fortissimumque, quia nullo nititur. Quocirca hoc vere unum, in quo nullus numerus, nullum in eo aliud praeterquam id quod est. [...] Deus vero a Deo nullo differt, [...]. Ubi vero nulla est differentia, nulla est omnino pluralitas, quare nec numerus; igitur unitas tantum".
14 See what is stated about Eriugena's interpretation of *Wisd.* 11, 21 in Chapter 3 of this book. On the influence of Boethian treatises on Eriugena's work, see d'Onofrio, "Giovanni Scoto", 711–720.
15 *Periphyseon*, III, 655B, 53; Sheldon-Williams, 109–111: "If unity is a unity of numbers, there never was unity without the numbers of which it is the unity. Also, if the numbers *flow* forth from the monad as from some inexhaustible source and, however much they are multiplied, come to an end in it, they would surely not be *flowing* forth from it if before their *flowing* forth they had not subsisted in it as in their cause". My italics.
16 Victorius, *Calculus*, *Praefatio*, 3: "Unitas illa, unde omnis numerorum multitudo procedit, quae proprie ad arithmeticam disciplinam pertinet, quia vere simplex est et nulla partium congregationem subsistit, nullam utique recipit sectionem". The relative pronoun (*quae*) in the first subordinate clause is ambiguous: the subjects to whom Abbo and Victorius refer it seem to be different. Anticipating what I will suggest in the chapter, according to Victorius, the relative *quae* is to be referred to the multiplicity (*multitudo*), whereas according to Abbo, to unity (*unitas*); so that for Victorius, multiplicity turns out to be the subject of arithmetic, while for Abbo, it is unity that stands as the subject of this discipline.
17 *In Calculo*, II, 4, p. 67. See the last section in Chapter 3, where it is show that Abbo introduces a distinctive explanation for the fact that created substances are not essentially good. Unlike Boethius, he justifies this assumption based on free will: if beings were substantially good, they would act rightly by an inner necessity, not by their free choice. Only God can be deemed substantially good since in Him there is neither the possibility nor the will to sin.
18 *Trin.*, II–III, 169–170. Expanded quotation is reported below in the main text of this chapter.
19 *In Calculo*, III, 1, pp. 72–73: "Cum omnium subsistentium quaedam sit principalis causa simpliciter, quicquid boni ab illa procedendo creatis prodest, hoc ipsi substantialiter permanet. Nec tamen materialem profitemur summae bonitatis originem ex qua omnia, per qua omnia, in qua omnia, sed eius participatione dicimus bona omnia. Nam divina substantia semper caruit et caret materia, licet non careat forma quam dum inprimit veluti sigillum rebus singulis, eas sine sui augmenti aut detrimenti mutabilitate procul dubio a se differre facit".
20 For Gilbert of Poitiers' 'universal singularism', see Valente, "Concreti", 110–115.
21 Abbo's conception of individuals and their link with shared formal traits (*dividua*) and God has been interpreted differently by Eva-Maria Engelen, whose reading leads to dissimilar conclusions compared to those presented in this book. Engelen based her understanding of individuals on the definition of the biblical number, measure and weight given by Abbo in the *Tractatus*, according to which these three are simultaneously coeternal, and each (*individua*) and all of

them corresponds to the one and only God (*In Calculo*, II, 13, p. 70: "tria simul aequiterna semper individua ubique et ubicumque tota unus deus sunt"). Therefore, according to Engelen, individuals consist in a type of entities or 'forms' such as number, measure and weight, which are eternal principles valid everywhere and always. From these kinds of individuals derive the respective *dividua*, which, in Engelen's view, are exemplified by the series of numbers and the scale of measurement. See Engelen, *Zeit*, 30–31: "An dieser Stelle für Abbo einen Individuenbegriff ein, der eng an den der Form, also der Idee, angelehnt ist, der in der Philosophie in dieser Verwendungsweise aber unüblich ist. Individua sind für Abbo ewige, von Gott erschaffene Prinzipien, die überall und immer Gültigkeit besitzen. Er erlautert seine Auffassung von Individua am Beispiel von Maß, Zahl und Gewicht. [...] Aus den Individuen wie Maß, Zahl und Gewicht stammen die Zahlen 2, 3, 4, ..., der konkrete Maßstab und das konkrete Gewicht. Diese existierenden Entitäten nennt Abbo wie Chalcidius "dividua"; erst durch die Sammlung sich ähnelnder dividua kann der Mensch mit Hilfe der Wissenschaften Arten bestimmen und Theoreme aufstellen". However, Engelen's position regarding individuals seems to me not to match Abbo's conception of number, measure and weight, as well as that of the concrete individuals and their relationship with both *dividua* and universals. While it is clear that the biblical number, measure, and weight determine the created reality, each one individually carrying out its own function, they are intrinsic to the divine essence, and therefore, all and each of them has to be considered as the one and only God – in this sense they may be seen as individual. As mentioned in the previous chapter, number, measure, and weight are conditions of enumerability, measurability, and being weighable, and neither the numerical series nor the scale of measurement properly derive from them. Engelen's conception of *dividua*, moreover, cannot dovetail with the concept of *dividua* presented by Abbo: multiple concrete individuals underlie single forms, and from these concrete individuals it is possible to derive shared traits (*dividua*) that lead to the formation of universals.

22 *In Calculo*, III, 1, p. 73: "Et haec differentia pro uniuscuiusque natura, quoniam formis, quas Plato hydeas vocat, plura subiacent individua. Ubi vero pluralitas, illic et numeri quantitas. Ab individuis quippe nascuntur dividua, ex quibus fiunt universa, per quae et collectio similitudinis, ut fiant species et theoremata per singulas artes".

23 *In Tim.*, 342, pp. 334–335. I would like to thank Christophe Grellard for this helpful insight.

24 *2 In Isagogen*, I, 11, p. 166. Cf. De Libera, *L'art*, 224–229; and Marenbon, "Boèce". I analysed the issue of universals and how they are intertwined to the epistemology of mathematics in Boethius' commentaries on the *Isagoge*, in Crialesi, "The Status".

25 *In Calculo*, III, 6, p. 77. Cf. *2 In Isagogen*, II, 7, p. 195. The labels 'metaphysical', 'physical', and 'logical' corresponding to the three senses of 'individual' are not mentioned by Abbo nor by Boethius. Rather, they stem from Gracia's reading of Boethius's *Second Commentary to the Isagoge*; see Gracia, *Introduction*, 74–75.

26 *In Calculo*, III, 1, p. 73.

27 *In Calculo*, III, 1–2, pp. 73–74: "Quippe unumquoddam individuum est omnium finis et initium. Unde Plato in Thimeo pro generatione animae aptam figuram repperit, in qua singularitas cacumini superimposita summitatem atque arcem obtinere consideratur, ut per eam, velut emissaculum quoddam, tanquam e sinu fontis perhennis providae intelligentiae quasi quidam largus amnis effluerет, ipsaque singularitas mens sive intelligentia sive Deus intelligi potest. Cum enim sit origo numerorum, omnibusque ex se substantiam subministret, rationesque eorum tam simplicem quam mutiplices ipsa contineat, ceteris numeris incrementis inminutionibusque mutatis atque ex propria natura recedentibus, sola inconcusso

iure est atque in statu suo perseverat semper eadem, semper inmutabilis et singularitas semper, quemadmodum divina omnia, quae nulla temporis progressione mutantur. Quidnam firmum existeret, si fundamentum titubaret? Geometrae nempe notam appellant, id est signum vel punctum, quod eius vicem obtinet nullis partibus distributum, propertereaque sub nullos sensus venit; est tamen et cernitur ratione animi. Ex qua nota producitur linea, ut ex unitate dualitas, licet sequentia perdant priorum potentiam, quia nec linea signum, nec duo unum. Idque manifestat praedicta psicogoniae figura, arithmeticae, geometricae, musicae et astronomiae subtilitatibus contenta ac hoc modo mirabiliter expressa". It is convenient to report also Calcidius' words, in order to see the resemblance between the two texts. *In Tim.*, 32, pp. 81–82: "Partitio consideratio est virium ordinatioque veluti membrorum actuum eius officiorumque et omnium munerum designat congruentiam tribus hoc asserentibus praecipuis disciplinis, geometrica arithmetica harmonica, ex quibus geometrica vicem obtinet fundamentorum, ceterae vero substructionis. Etenim quod nullas partes habet propertereaque sub nullos sensos venit, est tamen et animo cernitur, geometrae notam appellant, lineam vero sine latitudine prolixitatem, quae in notas suas desinit"; and *In Tim*, 39, pp. 88–89: "Nunc praestanda est ratio formae iustius triangularis, in qua sunt limites septem et sex intervalla duplicis et triplicis quantitatis. Nullam dico esse aptiorem figuram quam est haec in qua singularitas cacumini superimposita summitatem atque arcem obtinere consideratur, ut per eam, velut emissaculum quoddam, tanquam e sinu fontis perhennis providae intelligentiae quasi quidam largus amnis effluereet, ipsaque singularitas mens sive intelligentia sive Deus intelligi potest. Cum enim sit origo numerorum, omnibusque ex se substantiam subministret, rationesque eorum tam simplicem quam mutiplicatas ipsa contineat, ceteris numeris incrementis inminutionibusque mutatis atque ex propria natura recedentibus, sola inconcusso iure est atque in statu suo perseverat semper eadem, semper inmutabilis et singularitas semper, quemadmodum divina omnia, quae nulla temporis progressione mutantur".

28 *In Calculo*, III, 30, p. 91: "Unde liquet eandem unitatem numerabilium propagatricem non solum substantiam sed 'yperusian' dici, quod supersubstantialem appellant Latini". Abbo's knowledge of Eriugena's thought is clear from the letter that Fulbert of Chartres, disciple of Abbo, addressed to his own master: in this letter, Fulbert recalls the theory of *deificatio*, together with the distinction between the 'things that are' (i.e., the mutable things) and the 'things that are not' (i.e., the intelligible things). See Fulbert of Chartres, *Epistula* 1, p. 2. Cf. Jeauneau, "Fulbert".

29 *In Calculo*, III, 82, p. 122: "Ex uno aequalitatis fonte omnes inaequalitatis rivulos paulatim novimus profluere".

30 I will come back to this arithmetical procedure in the next chapter.

31 For the comparison between *unitas* and *punctum* cf. *In Tim.*, 34, p. 83: "Apex ergo numerorum singularitas sine ullis partibus, ut geometrica nota; cuius duplum linea"; and *In Somn.*, II, 2, 8, p. 9: "His geometricis rationibus applicatur natura numerorum, et monas punctum putatur, quia, sicut punctum, corpus non est, sed ex se facit corpora: ita monas numerus esse non dicitur, sed origo numerorum". For the definition of the *punctum*, see *In Tim.*, 32, p. 82: "Etenim quod nullas partes habet propertereaque aub nullos sensus venit, est tamen et animo cernitur, geometrae notam appellant"; and *In Somn.*, I, 16, 10, p. 92: "Punctum dixerunt esse geometrae quod ob incomprehensibilem brevitatem sui in partes dividi non possit, nec ipsum pars aliqua, sed tantummodo signum esse dicatur".

32 A definition of diagrams as conceived in medieval times is offered by Wallis, "What a Diagram". According to Wallis, one can think of a diagram as a 'visual argument', for it is not a mere illustration since it enriches and expands the contents of the text in which it is inserted, somewhat replacing the text itself. The diagrams reported in scientific or theological works can consist of graphic

elements, oftentimes of geometric type, and can incorporate textual inscriptions. For cosmological diagrams, see Obrist, "Wind"; and Obrist, "Démontrer". For diagrams in the Carolingian times, see Kühnel, "Carolingian Diagrams", where the scholar describes the diagram as a 'verbal imaging' whose aim is to facilitate the reader's access to the textual contents.

33 In this respect, see Michel Huglo's review of Peden's edition: Huglo, "Compte rendu".

34 Another Latin translation of the *Timaeus*, which was used by Augustine, is the one made by Cicero, dating back to 45 BC. It includes section 27d-47b of the *Timaeus*, yet with some gaps in the middle. In Cicero's translation, however, no lambdoid diagram is found. On this, see Hoenig, *Plato's Timaeus*, 44–55.

35 Plato, *Timaeus*, A 35b–36b. See Brisson, *Le même*, 35–55.

36 Del Forno, "The mathematical structure". Cornford, *Plato's Cosmology*, 66. While Del Forno stresses that the demiurge makes a division into portions, Cornford assumes that the demiurge firstly formed a long strip from the psychic compound, which he then divided into harmonic intervals. It is worth precising, however, that, in Plato's dialogue, the passage of the dialogue concerning the division of the soul into strips or strings (36b7 ff.) follows the passage concerning the section of the soul into parts according to numerical ratios.

37 For a detailed explanation of the proportions structuring the World Soul, see Panti, *Filosofia della musica*, 21–24; and Handschin, *The Timaeus scale*. For music with respect to Plato's thought, see Pelosi, *Plato on Music*.

38 This aspect was already acknowledged by Proclus (*In Platonis Timaeum*, II, 199). Regarding the discrepancy between the 'noetic harmony' and the 'phenomenal harmony' with respect to Plato's World Soul, it is worth quoting Hicks, *Composing the World*, 197: "Soul has only a noetic harmony: it is not corporeal (even if the Demiurge manipulates it as if were – mixing it, cutting it, shaping it); it has no dimension (even if the Demiurge measures it and divides it as if it did); and it is not quantitative (even if the Demiurge imparts to it a distinctly quantitative structure). Phenomenal harmony, however, is all of this – corporeal, dimensional, and quantitative".

39 Commentaries on the *Timaeus* often took the form of thematic monographs, focusing on some sections of the *Timaeus*. The choice of the sections to be commented on somewhat reveals the philosophical inclination, so to speak, of the author and their time. This is the thesis of Ferrari, "I commentari"; and Ferrari, "Struttura".

40 Plutarch, *De animae procreatione in Timaeus*, 29, 1027c-d. Along with the lambda diagram, Plutarch recalls another diagram, namely, the linear one, which he traces back to Theodore of Soli. According to Proclus (*In Platonis Timaeum*, II, 171), the linear diagram was also adopted by Porphyry. Macrobius refers to the same linear diagram in *In Somn.*, II, 2, 17, pp. 11–12.

41 Adrastus' work is lost. However, many of his reasonings and interpretations can be retraced via Theon of Smyrna. Especially concerning the lambda diagram, see Theon of Smyrna, *Expositio*, 95, pp. 145–146.

42 *In Tim.*, *Ad Osium epistola*, p. 6. For an overview of Calcidius standing before Plato's *Timaeus*, see Reydams-Schils, *Calcidius*.

43 *In Tim.*, 2, p. 58. The reference to the clarifying contribution of the technical notions is then reiterated at the end of the first part of the commentary, *In Tim.*, 119, p. 164.

44 *In Tim.*, 50, pp. 99–100.

45 Figures no. 7, 8, and 9 in *In Tim.*, respectively on pp. 82, 90, and 98. I refer to two studies dedicated to the diagrams and their analogical function in Calcidius' commentary: Somfai, *Calcidius' Commentary*, especially 213 for the lambda diagram; and Bakhouche, "Lectures musicales". For Calcidius on music, also

with reference to lambda diagrams, see Meyer, "L'âme du monde"; and Huglo, "L'étude des diagrammes". Finally, for Calcidius' references to Pythagoras's authority in his *Commentary*, see Hoenig, "Calcidius", 272–273.
46 *In Tim.*, 33, pp. 82–83; and *In Tim.*, 45, pp. 93–94.
47 *In Tim.*, 35, pp. 75–76. Cf. also *In Tim.*, 40–50, pp. 89–100.
48 *In Somn.*, II, 2, 19, p. 12. For an in-depth study of music theory in Macrobius, see Meyer, "La théorie". In the musical section of the second book of his *Commentary*, Macrobius indicates with *hemiolium* the sesquialter arithmetic ratio, while Calcidius refers to it by calling it *sescuplus*. The sesqui-third, on the other hand, is called *epitritus* by both. Besides this difference, Calcidius and Macrobius share the same musical terminology.
49 *In Somn.*, I, 6, 2, pp. 24–25.
50 *In Somn.*, II, 2, 14, p. 10. A cubic number can be associated with a solid if one considers that a line is contained by two points, a surface by four and a solid by eight. A similar observation occurs also in *In Tim.*, 33, p. 82.
51 *In Somn.*, I, 6, 46–47, pp. 35–36. The translation, which has been slightly modified, is taken from Macrobius, *Commentary*, 109. Macrobius' reasoning is also present in *In Tim.*, 38, p. 87. As we will see, Abbo, too, will dwell on the number seven and yet not with respect to the World Soul, but rather to the regularity of some natural phenomena. I will deal with it in the next chapter.
52 Different is the opinion of Flamant, *Macrobe*, 325–326, according to whom Macrobius actually used only the linear diagram (introduced by Theodore of Soli) when he happened to offer a deeper explanation of the Timaean psychogony in his commentary (*In Somn.*, II, 2, 17, pp. 11–12).
53 Macrobius's commentary includes five diagrams of astronomy and cosmography in total: Barker-Benfield, "Macrobius", 222–232.
54 Huglo, "La réception", 15–16.
55 The manuscript Paris, BnF, lat. 6370 is the oldest witness of Macrobius' work, and is glossed by Lupus of Ferrières and Heiric of Auxerre. Paris, BnF, lat. 16677 was originally copied at Fleury (Mostert, *The Library*, 236, BF1221-1222). Cf. Caiazzo, "Mains célèbres", 173–174.
56 For the manuscripts of Macrobius' commentary from Fleury, I refer to Barker-Benfield, *The Manuscripts*, vol I, 87–112. The migration of the lambda diagram from the margin into the text is not an issue taken into account in Peden, *Glosses*. Peden points out the presence of the lambda diagram, accompanied by glosses and placed in the margin of the specific passages of the *In Somn.*, namely, at I, 6, 2–5 and II, 2, 16–17. See Peden, *Glosses*, 106–108, vol. I.
57 Huglo, "La réception", 11 13 and 16. Cf. section 2 in Chapter 3, especially note 41.
58 In fact, the works cited were originally preceded by Ambrose's *De fide*, which is now found in manuscript Paris, BnF, lat. 1747, as argued in Caiazzo, "Abbon de Fleury", 16–17.
59 *In Calculo*, III, 3, p. 74.
60 *Musica*, I, 10, pp. 197–198. Cf. *In Somn.*, II, 1, 8–13, pp. 4–5.
61 Cf. manuscripts Berlin, Staatsbibliothek, Phill. 1833 (Rose 138), fol. 9v; and Karlsruhe, Badische Landesbibliothek, 504, fol. 92v. In manuscript Wien, ÖNB, 2269, fol. 135r, the diagram is devoid of musical references, but is flanked by the second diagram presented in Calcidius' *Commentary* (Figure no. 8, *In Tim.*, p. 90). Cf. Huglo, "D'Hélisacar", 223.
62 Huglo, "Compte rendu", 225–226.
63 *Musica*, I, 3, p. 191: "Est enim consonantia dissimilium inter se vocum in unum redacta concordia".
64 All arithmetical ratio mentioned here are explained in Chapter 6. It seems none the less convenient to briefly clarify the terminology here used. Between two numbers, the sesquialter is the greater one that contains the smaller and half of the

smaller. For example, given a ratio $a : b$, where a is the greater and b the smaller, a is sesquialter if $a = b + {}^1/_2\, b$. The same thing for the sesqui-third, which, given two numbers, corresponds to the greater one that contains the smaller and its third. For example, given a ratio $a : b$, where a is the greater and b the smaller, a is sesqui-third if $a = b + {}^1/_3\, b$.

65 Panti, *Filosofia della musica*, 97–101.

66 *In Calculo*, III, 2, pp. 73–74: "Idque manifestat praedicta psychogoniae figura, arithmeticae, geometricae, musicae et astronomiae subtilitatibus contenta ac hoc modo mirabiliter expressa". Differently, Calcidius mentions only arithmetic, geometry and music.

67 See the previous section in this chapter, especially notes 27 and 31.

68 Peden (*In Calculo*, 74, note 29) defended the first of these two views.

69 *In Somn.*, II, 3, 12, p. 17.

70 *In Somn.*, II, 3, 14, p. 17. See the *nota ad locum* by Armisen-Marchetti, *In Somn.*, 108–109; and Flamant, *Macrobe*, 371–373. Macrobius' proportion can be expressed as follows: T-So:T-L=2:1; T-V:T-So=3:1; T-Me:T-V=4:1; T-Ma:T-Me=9:1; T-I:T-Ma=8:1; T-Sa:T-I=27:1. Sigla stand for the astronomical bodies' Latin names.

71 *In Tim.*, 96, p. 148. Calcidius' proportion can be expressed in the following way: T-L = 1; T-So:T:L=2:1; So-Ve:T-L=3:1; V-Me:L-So=4:2; Me-Ma: So-V=3:1; Ma-I:T-L=9:1; I-Sa:T-L=27:1.

72 Abbo, *Dulce ingenium musicae*, 34: "Hinc nervi, id est cordae artificalis musicae caelesti musicae comparantur. Nam ypate meson saturno ceterique septem planetis, quamquam diverso ordine secundum Boetium atque Macrobium". Cf. *Musica*, I, 27, p. 219. On the music of celestial bodies in Boethius, see Panti, *Filosofia della musica*, 94–95. Shortly after this sentence in the *Dulce ingenium musicae*, Abbo reports word by word Macrobius' commentary, that is *In Somn.*, II, 3, 11, p. 17: "Iure igitur musica capitur omne quod vivit, quia caelestis anima, qua animatur universitas, originem sumpsit ex musica".

73 An alternative interpretation of unity within a strictly astronomical context could be found in Abbo's acrostic poem, the *Ephemerida*, particularly in the reference to the *unius ordo* (Baker and Lapidge, "More Acrostic Verse", 14, l. 23). Nadja Germann has delved into the meaning of this *unius ordo* in her attempt to discern the nature of the poem, which she suggests is astronomical rather than linked to computus. Germann posits that, in the *Ephemerida*, Abbo aims to celebrate the 'sacred things' ("ardua sacraria" at l. 1; "Alme deus, te nosse mihi concede fideli" at l. 29), and in this context, she identifies a divine order in the *unus*. According to Germann, this *unius ordo* ensures the perfect regularity of the alternation of lunar and solar courses. Abbo warns that, although the Sun and the Moon have different speed, there is an 'order of the one' which guarantees the their compliance. This perfect 'order of one' rules over the cosmos and, therefore, would be the image of God himself. However, it is important to note that while this reading might seem to extend Abbo's henology to include astronomy, such a conclusion could be premature. The sentence in which the so-called 'order of the one' occurs is the following: "Hic semper servus haec est velotior etsi / ambos aequali cursu ciet *unius ordo*" (vv. 22–23, mine italics), whose translation is: "This [i.e., the sun] is slow, <while> this [i.e., the moon] is faster, despite the order of the one [i.e., the latter] regulates both <the heavenly bodies> according to a fair course <for each of them>". The appropriate speculative context for understanding the *unius ordo* is neither theological nor henological but rather purely astronomical. The reference to the Moon's course indicates its role in marking the duration of the solar course, with the synodic month representing the time for the realignment of the Moon, Sun, and Earth. I express gratitude to Cecilia Panti for discussing the ambiguity in this passage with me. See Germann, *De temporum ratione*, 171–176; Germann, "À la recherche", 166–172.

74 *2 In Isagogen*, III, 7, pp. 221–223, especially p. 223.
75 *2 In Isagogen*, III, 7, p. 224.
76 *2 In Isagogen*, I, 10, pp. 161–162.
77 Boethius, *Contra Euthychen*, 4, p. 220, ll. 296–300.
78 The Boethian origin of 13th-century logical and theological theories on the so-called 'transcendentals' are highlighted in Valente, "Names"; Valente, "*Ens, unum, bonum*"; Valente, "*Illa quae transcendunt*". On transcendental terms, see Aertsen, *Medieval Philosophy*; Aertsen, "*Transcendens-Transcendentalis*".
79 See also d'Onofrio, "Giovanni Scoto".
80 Rand, *Johannes Scottus*, 68.
81 Cf. Marenbon, *Early Medieval Philosophy*, 82.
82 *In Calculo*, III, 3, pp. 74–75: "Qui investigans naturam unitatis praetendit rationem mundani principii. Dicendumque inprimis, quare ipsa unitas sectionem non recipit. Omne quod est, unum est, et quicquid unum est, illud sit necesse est. Sunt nimirum unum et est ad sua invicem praedicationem convertibilia. Quae autem in sua praedicatione convertuntur, aequalia sunt. Aequalia sunt igitur unum et est, quia nec unum dici potest quod non est, nec esse quod non unum, quanquam sint quaedam rerum universalia quae, cum in pluribus uno eodemque tempore totum sint, tamen dicuntur esse, ut 'homo' qui uno eodemque tempore in Socrate et Platone totus est, ex quibus suum esse colligit. Quo fit ut 'homo' intellectus universaliter non sit, qui semper particulariter causa unitatis existit".
83 This is suggested by Valente, "*Ens, unum, bonum*", 494.
84 *In Calculo*, III, 4, p. 75: "Sciendum tamen quod oporteat in numerando differentiam intellegi, quia aliud est quo numeratur, aliud quod numeratur. Quo numeratur quidem est numeri vocabulum, ut unitas; quod vero numeratur a vocabulo denominatur, ut ab unitate unus. Idem modus in duobus et dualitate tribusque ac trinitate et reliquis. Nec unquam alicuius rei vocabulum recipit augmentum aut detrimentum, cum tamen quaedam denominata in conparatione gradus veniant. Servantque illa quodammodo primitiva accidentium regulam, quae, sicut per se non subsistunt, ita nec per se aliquid pati possunt. Cum itaque dualitas per se sectionem refugiat, quanto magis unitas? Sed unitatem primae substantiae, in quae cogitatio dividat, non invenit, dualitatem vero vult quidem, sed per se nequit. In his igitur quae numerabilia sunt et duorum facilis partitio et unius quoquo locorum sectio, ita tamen ut si verbi gratia sit unus pes, aut aliquid talium mensurae vel ponderis, ei secando nec inferri nec auferri quid possit, nisi mutando qualitatem prioris unitatis. Nec hoc asserimus quod uni librae aut pedi addi aliquid non possit, sed quod unum nec aumento nec detrimento eget, ut aliquid unum sit". *Substantia prima* recalls the Aristotelian notion of 'first substance' (*Cat.* 2a11-3a7); in this context, then, the unity of 'first substance' is the word that is said of something individual and concrete.
85 *Arithm.*, I, 3, 2, p. 12. On multitude and magnitude, cf. *Arithm.*, I, 1, 3, pp. 6–7.
86 Victorius, *Praefatio*, in *In Calculo*, 3. On Victorius' account of unity see Chapter 2.
87 *In Calculo*, III, 7, p. 77: "Sed quia omnium quae sunt, alia sunt concreta et sibi herentia, ut homo, lapis; alia disgregata et suis partibus determinata, ut grex, populus; idcirco additum est post unitatem, 'unde omnis multitudo numerorum procedit'. [...] Discretis enim et a se divisis praeest multitudo, sicut continuis et sibi herentibus magnitudo, quia et multitudo ab uno in inmensum crescit et magnitudo in infinitum decrescit. Semperque vocabulo multitudinis significatur collectio pluralitatis et appellatione magnitudinis uniuscuiusque quantitas corporis [...]. Cumque numerus sit unitatum collectio, et quodammodo multarum partium in unum redacta coacervatio, recte de eo dictum est quod sit multitudo".

88 *Arithm.*, I, 1, 4, p. 7. On the division and order of these disciplines, see Guillaumin, "L'ordre".

89 Victorius, *Praefatio*, in *In Calculo*, p. 3.

90 *In Calculo*, III, 8–9, pp. 77–78: "Ad geometricam nanque et ad musicam et ad astronomiam unitas pertinet, sed proprie, sicut ex ea factus numerus, ad arithmeticam, quae dicitur numerorum mensura. Mensuratur autem numerus aut sola unitate, quae est omnium communis mensura ut dictum est, ut ternarius, quinarius; aut uno quolibet numero sua quantitate multiplicato, ut quaternarius binario, et novenarius ternario; aut non solum uno sed etiam duobus, ut XVI non solum quaternarium per se sed etiam binario et octonario. [...] Sunt tamen qui pluribus mensuris constant, ut XII prima <mensura> perficitur unitate, secunda senario, sexta binario, tertia quaternario, quarta ternario. [...] Sunt etiam numerorum alii commensurabiles, alii incommensurabiles. Et commensurabiles quidem sunt qui una qualibet communi mensura utuntur, ut VIIII, XII, XV. [...] Incommensurabiles vero sunt qui excepta unitate nullam communem mensuram recipiunt, ut VIII et VIIII".

91 Marziano Capella defines 'part' as an integer number that produces a greater number by adding another unit (or 'part'). For instance, 7 corresponds to seven units (1+1+1+1+1+1+1) that are seven 'parts'; in the same way, the 3 and the 2, that added to the 1 produce the 7, are considered as 'parts'. Martianus, *De nuptiis*, VII, 752, p. 21: "At singula et si qui etiam solidi numeri immixti singulis inserentur, partes appellabo, ut in VII vel totidem singula vel etiam bis terna singulo adiecto".

92 *In Calculo*, III, 10, pp. 78–79: "Saepe enim lineae aut alterius rei quaelibet pars exquiritur a minus peritis [...] et dum eam singillatim unitatibus discutient interdum frustrato labore deficiunt. Quod ne fiat, prius investigandae sunt partes numeri ad cuius rationem corpus dividere conaris. Sciendumque quia quot divisiones suscipit ille numerus, totidem recipit illi numero subiectum corpus, ut verbi gratia lineae exploratur pars duodecima. Sumpto itaque, si forte necesse est, circino, eius alterum pedem lineae capiti, alterum medio figis et ita duodenarii sex unitates hinc et sex inde contemplaberis. Quarum sex unitatum medietatem circino explorabis itidem, inventoque ternario, qui se dividi pro eo quod impar est, non sinit, eum circino rursus distribues in unitatibus singulis. Igitur tertia pars huius ternarii secundum quantitatem permensi corporis, duodecima est duodenarii".

93 The procedure of the division by half is introduced in Boethius' *De institutione musica*. On the basis of Ptolemy's *Harmonica*, Boethius presents the way human senses operate in the division of a segment. In Boethius' opinion, this way of proceeding proper of the senses is more inaccurate than the one adopted by reason. For the latter is able to operate more precise divisions, for instance, not only by half, but also by eight, and so on. Cf. Panti, "Boethius", 19–22.

94 *In Calculo*, III, 11, p. 79: "Per medietates imparis fit tali modo: medietas XVII est VIII et semis. Cuius medietatis est rursus medietas IIII et quadrans; huius quoque medietas II et sescuncia, et huius medietas I cum semuncia et sicilico. Sicque unitas imparis numeri, quae eius aequali divisioni inpedit, ad aequalitatem in corpore redigitur unciatim; nunquam tamen per medietatis sectionem ita simplex exit".

95 *In Calculo*, III, 12, p. 80: "Numeros ergo dum dividis aut sibi conparas, nonnihil agere in corpore te credas, nisi quando pro argumento accipis ad repperiendum quodlibet spatium temporis. Quod vero seu integrum corpus sine incisione partiatur, seu etiam cum incisione, minor numerus maiores partes corporis conplectatur et maior minores, etiam pueri noverunt, qui famelici magis diligunt tertiam partem tortae panis quam quartam et quartam quam octavam. Quotienscumque nimirum aliquid partitus fueris incidendo, procul dubio numeri pluralitas crescit more multitudinis, quantitas vero corporis decrescit instar magnitudinis".

96 Cf. Abbo, *De cursu*, 126.

5 The Composition of Reality

5.1 Natural Compounds

Abbo introduces the concept of composition by delineating a distinction between natural and artificial compounds. The former category pertains to entities endowed with at least one of the three powers of the soul – growth, sensation, and discernment – while the latter encompasses products of human will and skill, achievable, for example, by applying the principles of a liberal discipline.[1] Abbo's objective is to demonstrate that the developmental trajectories of both these compound types adhere to a dual-phase structure: initially, they attain the zenith of their growth, achieving peak complexity, and subsequently, they undergo corruption (in the case of natural compounds) or reduction (in the case of artificial ones), reverting to their original states by retracing their developmental steps in a backward fashion. In the context of natural compounds, these two developmental stages exhibit an inverse proportionality, wherein the growth trend is inversely proportional to that of corruption. As elucidated below, the same proportional relationship holds true for the development of artificial compounds.

> After having discussed the simplicity of unity, [Victorius] proceeds in a reasoned order to discourse on composition opposite to it [*scil.* to the unity], which is either by nature or by will. But things that are composite by nature, if they take their increase in the legitimate progression, can in no way be dissolved otherwise than they were composed. Neither the motion of increase nor the natural progression take any other path to decrease than that from which it grew.[2]

Abbo illustrates the progression and regression of natural compounds through two case studies: the temporal evolution of human physical strength and the variation in the Moon's magnitude throughout its phases, which are represented in Figures 5.1 and 5.2, respectively.

These examples appear to be intentionally drawn from both the microcosmic and macrocosmic realms, possibly to underscore a shared rationale between them. In these specific instances, this rationale manifests through a

DOI: 10.4324/9781032643472-6

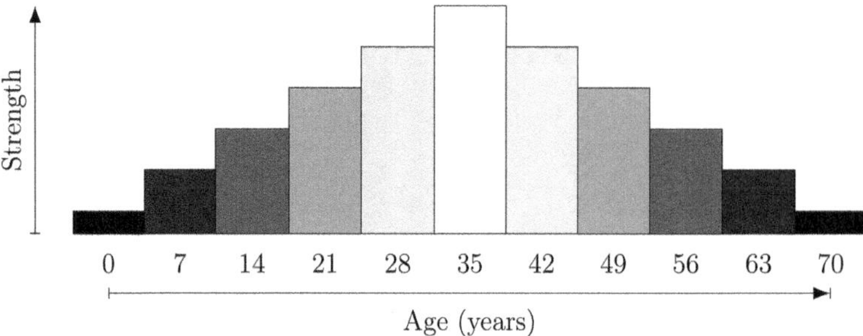

Figure 5.1 Progression of human strength and age.

shared scheme based on number seven. The development of human strength and lunar phases adheres to a standard division into modules of seven, corresponding to years and days, respectively. Notably, both phenomena exhibit an inversely proportional trend in their evolution. Human vigour experiences an incremental rise until reaching its zenith in the fifth seven-year cycle, that is, at thirty-five years. Subsequently, a decline ensues, intensifying at intervals of seven years, culminating in the tenth seven-year cycle, that is, at seventy years, when strength diminishes entirely, returning to the initial degree present at conception. Similarly, the moon follows a pattern where it is fully visible at the pinnacle of its synodic course, in the second seven-day cycle, or on the fourteenth day. Thereafter, it progressively diminishes in visibility each week until it is completely obscured on the fourth week, that is, on the twenty-eighth day. This parallelism in developmental cycles, governed by the number seven, reveals the interconnected rationality between the microcosm of human life and the macrocosm of synodic cycle as conceptualised by Abbo.

> What is evident according to physiologists, who distribute the course of human life from the very time of conception through periods of seven days, months, and years, is that from the completed fifth set of seven years, that is, at the age of thirty-five, human strength reaches its peak from the day of birth, and likewise, it recedes in the same intervals, and at the seventieth year, that is, within a lapse of other five sets of seven years, it declines as bending backward. The same is experienced in the waxing and waning of the moon, which, being fuller in the second week, inversely starts waning over the next two weeks, that is, on the twentieth day.[3]

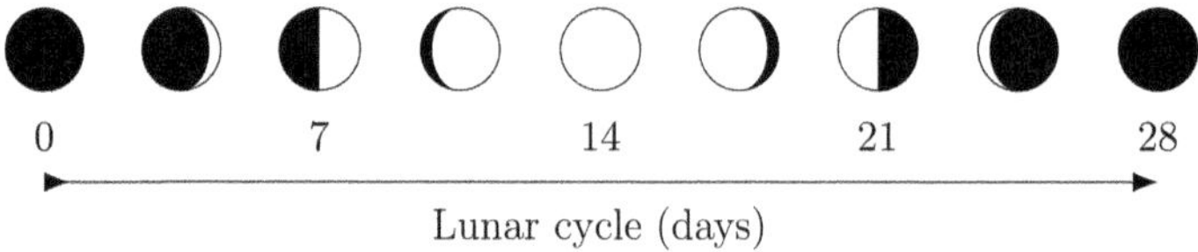

Figure 5.2 Lunar cycle.

Abbo explicitly attributes his comprehension of the growth and decline of natural compounds to the perspectives articulated by 'physiologists'. Behind such epithet, one might discern figures like Calcidius, Macrobius, and Martianus Capella, based on the significance they attribute to the number seven within the natural realm. Notably, they highlight the correspondence of the number seven with various anatomical features, such as the seven holes in the human head, internal organs, body parts, and pathological processes.[4] In this respect, Macrobius stands out for projecting the significance of the number seven into both biological and astronomical domains. In his *Commentary*, he examines the span of human life from embryonic development up to seventy years, whereas Calcidius and Martianus limit their considerations to the fifth seven-year cycle (i.e., thirty-five years). Macrobius contends that the biological milestone of seventy years represents the ideal state for human life, since individuals who reach this age are liberated from all responsibilities and can exclusively devote themselves to the pursuit of wisdom.[5] Moreover, Macrobius delves into the lunar aspects more extensively than Calcidius and Martianus. He attributes the number seven not only to the sidereal revolution, representing the time for the Moon to return to the same point in its orbit, but also to the synodic revolution, denoting the time for the Moon to realign with the Sun.[6]

It is noteworthy that neither Calcidius, Macrobius, nor Martianus explicitly reference two inversely proportional phases when addressing the unfolding of human life and the Moon's magnitude during a synodic cycle. Particularly in Macrobius' perspective, the development of human life is depicted as a linear process culminating in perfection only at its conclusion. Abbo, on the contrary, offers a distinctive viewpoint by accentuating the two symmetrically inverse phases in the development of natural compounds, as depicted in Figures 5.1 and 5.2. Abbo's exposition of the Moon phases, as previously outlined, is not intended to provide a detailed account of astronomical theories – a subject that he addresses in other works. Rather, Abbo's primary objective in commenting the *Calculus* is to demonstrate how natural compounds, whether celestial or terrestrial, undergo a gradual dissolution, retracing the same developmental stages in reverse order. His central aim is to illustrate that all these compounds share a common origin, a concept that carries theological and philosophical implications. The existence of an 'original status' for compounds is inferred from their cyclical return to their starting point. At the conclusion of its trajectory, the Moon is entirely obscured, reminiscent of its appearance at the outset. Similarly, at the end of a human life, physical vigour wanes, mirroring the null state of vitality at conception.

5.1.1 The Arithmology of Natural Compounds

Abbo's exploration of natural compounds extends to the symbolic or arithmological significance of some numbers that stand as important steps in the age of humans and the lunar cycle: 70 (equivalent to ten seven-year cycles) and 28 (equivalent to four seven-day cycles), respectively.[7] The symbolic significance of the number 70 emanates from the combined symbolic values of its factors, seven especially. Abbo regards seven as a number emblematic of wisdom – an idea reinforced by the mention of the seven columns of Solomon's temple, symbolising the seven liberal arts, as found in the *Tractatus de numero, mensura et pondere*.[8] In all likelihood, his interpretation stems from Calcidius, Macrobius, and Martianus, who articulated an arithmological discourse around number seven in the following way. Among the first ten natural numbers, seven stands out as the sole number that cannot be derived from any factor smaller or equal to ten – excluding unity, naturally – and cannot, through multiplication, give rise to any number smaller or equal to ten. To illustrate this distinction, let us contrast seven with another prime number, specifically, three. Similar to seven, three possesses unity as its sole factor; however, unlike seven, three can generate two numbers within the first ten numbers (i.e., six and nine) when multiplied either twice or by itself. Calcidius, Macrobius, and Martianus draw connections between these arithmetical characteristics of the number seven and a characterisation of the goddess Minerva. Minerva, according to their interpretations, is depicted as a virgin born from a single parent, namely Jupiter. Similarly, the number seven is considered 'virgin' in the sense that it does not give rise to any number smaller than ten and is 'born' in the sense that it can only be produced by multiplying the number one, which, within this framework, is identified as 'its only parent'.[9]

If the arithmological value of the number 70 derives from its factors – especially seven –, the specific significance of the number 28 is deduced from its mathematical property as a 'perfect number', showing clearly that Abbo's analysis is rooted in Boethius' *De arithmetica*. Boethius categorises numbers into three groups: 'perfect', 'imperfect', and 'more than perfect'.[10] A 'perfect number' is defined by its equivalence to the sum of all its factors. Notable examples are 6 and 28, where their respective sums are derived as follows: $1 + 2 + 3 = 6$; $1 + 2 + 4 + 7 + 14 = 28$. In contrast, 'imperfect numbers' cannot be generated by adding all their factors, resulting in a sum smaller than the number itself. Examples include eight ($1 + 2 + 4 = 7$) and ten ($1 + 2 + 5 = 8$). 'More than perfect numbers' are those for which the sum of factors exceeds the number itself, exemplified by 12 ($1 + 2 + 3 + 4 + 6 = 16$) or 18 ($1 + 2 + 3 + 6 + 9 = 21$).[11] In addition to these arithmetical considerations, Abbo introduces the notion that perfect numbers are metaphorically 'besieged' (*obsessi*) by both imperfect and more than perfect numbers. The use of the term 'siege' resonates with the terminology employed in rhythmomachy, a mathematical chess game precisely based on Boethian number theory. Abbo's reference to rhythmomachy gains credence considering that perfect numbers like 6 and 28 could be in fact envisioned as pawns on a chessboard, susceptible to the strategic manoeuvres of an adversary.[12]

The arithmological value of 28 finds significance also through one of its factor, namely, four. Since the time of ancient Pythagoreans, the number four has been deemed a 'sacred number', a designation bolstered by the fact that ten, another number symbolising perfection, can be derived by summing the first four natural numbers. However, the import of this number becomes particularly evident within the physical realm. Abbo accentuates the role played by the number four in establishing connections between qualities that harmoniously structure the world. It acts as a link for everything that is within the world, which appears somewhat "connected by the four qualities of harmonic sweetness" ("mundanae conexioni per quattuor qualitates armonicae suavitatis").[13] From this perspective, Abbo seems to suggest a strong interconnection between musical and physical harmony, positing that a foundational harmony underpins the physical universe.

To comprehend how the number four is proposed to structure the world, it is essential to elucidate both the physical and musical senses in which the number four can be understood. The conception of the number four as fundamental to the physical structure of the world can be traced back to Plato's *Timaeus*. There (31b-32c), Plato delineates how the demiurge establishes fire and earth as the foundational elements of the cosmos. Positioned at the upper and lower extremities of the universe, these elements, being three-dimensional bodies like all solid substances, necessitate connection through two 'middle terms', namely the intermediate elements of air and water. Calcidius expounds upon this Timaean doctrine in the section of his *Commentary* addressing what matter is (*de silva*).[14] In elucidating the transformation of the four elements – fire, air, water, and earth – into one another, Calcidius resorts to a set of qualities (*subtilis*, *corpulentus*, *acutus*, *obtunsus*, *mobilis*, *immobilis*) to be attributed to each of the four elements. Leveraging the pattern of a geometric proportion with four terms, Calcidius strategically allocates the six qualities among the four elements: the fire is thin, moving, and sharp; the air is thin, moving, and blunt; the water is blunt, corpulent, and moving; and finally, the earth is blunt, corpulent, and motionless. Distributing three qualities per each element, the extremes – fire and earth – are linked with each other, sharing at least one quality, and each also shares two qualities with the intermediate elements – air and water. This method of organisation ensures a balanced and harmonious arrangement of qualities, showing how it is possible to relate the four elements that are at the basis of the physical realm.[15] Macrobius, too, directs his attention to the same Timaean issue, basing his interpretation on the four Aristotelian qualities (hot, cold, wet, and dry). In this context, the two outer elements, namely fire and earth, and the two intermediate elements, air and water, are interconnected in what he terms a "iugabilis competentia" or a compatible pairing, in a sort of circular way. Specifically, fire is linked to air, air to water, water to earth, and earth to fire. Each element is connected to the next one through at least one of the two qualities inherent to it. Consequently, earth and water share the quality of cold, water and air share wetness, air and fire share heat, and, finally, fire and earth share dryness. This other type of interconnected arrangement illustrates a congruent relationship among the four elements, where each is bound to the next by virtue of shared qualities.[16]

The connection of worldly things suggested by the number four can arguably be understood in the context of the interplay among the four elements, each bound together by a specific set of qualities. This interpretation finds support in Abbo's depiction of God, where he employs a well-known verse from Boethius' *Consolation of Philosophy* to state that God "binds the elements by numbers" ("numeris elementa ligat").[17] The idea here is that the numerical structure plays a significant role in establishing the intricate connections and harmonies within the physical world, as reflected in the divine act of binding the elements through numbers. In evoking the physical relevance of the number four, Abbo likely alludes to the way elements can be linked through their qualities like in a mathematical proportion including more than two terms, as expounded by Calcidius and Macrobius.

On the other hand, it is crucial to consider also the 'harmonic sweetness' Abbo refers to the number four. The utilisation of the number four to convey the foundational music of the world undoubtedly resonates with the harmonic and mathematical structure of the World Soul. As previously discussed, Calcidius interprets the Timaean psychogony in the context of music theory. In this interpretation, Calcidius introduces harmonic ratios that form the basis of the Pythagorean musical scale. The ratios of the Pythagorean scale are precisely derived from the first four numbers: the fourth (4:3), the fifth (3:2), and the octave (2:1). This, coupled with the earlier considerations, underscores that Abbo is distinctly grounded in a Neopythagorean framework also in his understanding of the physical realm. The association of the number four with harmonic principles suggests a deep-seated understanding of the role of music theory in expressing the underlying harmony of the cosmos, integrating with Platonic and Neopythagorean doctrines.

5.1.2 Natural Compounds and Change: The Phases of the Moon

The primary characteristic of natural compounds lies in the fact that any alterations they undergo are not influenced nor controlled by human will. Abbo illustrates this principle using an astronomical example: the phases of the Moon. His line of reasoning unfolds as follows. A discernible external factor, responsible for the changes in the Moon's phases, can be identified. This factor is the Sun, serving as an illuminating source for the Moon. The Moon receives sunlight to the extent possible, resulting in a perpetual division between one darkened portion and another gradually illuminated section, dependent on the Moon's position with respect to the Sun.

> The globular shape of the moon gradually increases and decreases, [...] one part of it admitting brightness like a mirror, which it reflects back to the Earth, while the other, darker part receives no light. The spherical solidity of any half obscured object, such as an egg, expresses a resemblance to this phenomenon if you move the shadow just cast a little, so that you may observe the horned whiteness akin to the newly emerging

> moon. Continuing to turn it, you will see all the phases of the moon up to the full moon and again to the waning. This occurrence is evident in various lunar shapes, such as the crescent, half-moon, gibbous, and full moon, in both hemispheres.[18]

By practically explaining the lunar phases, Abbo not only clarifies an essential feature of natural compounds but also imparts basic astronomical notions to the reader and his students. To convey this, Abbo describes a practical experiment involving the projection of a shadow onto a spherical object – an egg will do –, which mimics the Moon's continuous cycle of illumination and obscurity. The shadow is gradually cast in a way that simulates different lunar phases, including the crescent Moon (*monoydes*), the first or third quarter (*diatomos*), the gibbous Moon (*amphicirtos*), and the full Moon (*panselenos*).

The terminology Abbo employs for the lunar phases is noteworthy. More specifically, the name assigned to the full Moon, *panselenos*, reappears only twice after Abbo's time: first in the 11th century in Byrhferth of Ramsey's *Enchiridion* and then in the 12th century in William of Conches' *Dragmaticon philosopiae*.[19] Moreover, Abbo's detailed terminology precisely matches the astronomical diagram found in his *Sententia de ratione spere*, which is presented also in both Byrhtferth's and William of Conches' works.[20] This unique continuity of the term *panselenos* and its association with the specific lunar diagram from Abbo's astronomical text, the *Sententia*, which is shared by subsequent authors, emphasises the enduring influence of Abbo's contribution to the understanding of lunar phases and their depiction. In other words, it stresses the significance of Abbo's work in shaping the astronomical terminology and conceptual framework adopted by later scholars in the Middle Ages. The quoted passage from the *Explanatio* can thus be viewed as an insightful commentary on the Moon phases diagram included in Abbo's *Sententia*. The astronomical diagram in question (Figure 5.3), takes the form of a circle, with the Moon phases depicted along its circumference, accompanied by their respective names. The diameter of this circle symbolises the 'earthly horizon', representing the viewpoint of a hypothetical observer. The full Moon is situated at the left end of the diameter, while the Sun, indicating the new Moon, is at the right end. Barbara Obrist highlights that Abbo's diagram, unlike previous ones depicting lunar phases, emphasises the significance of the line representing precisely the 'earthly horizon'. This visual element symbolises the spatial perception of the celestial phenomenon, taking into account the observer's terrestrial perspective.[21] In interpreting Abbo's passage from the *Explanatio* as a commentary on this astronomical diagram, one can appreciate his mention of two hemispheres ("utriusque hemispherii") – the visible and hidden terrestrial hemispheres from the individual observer's standpoint. This concise reference gains clarity when one considers the *Sententia* diagram, reinforcing the interconnectedness of Abbo's theoretical and visual explanations.

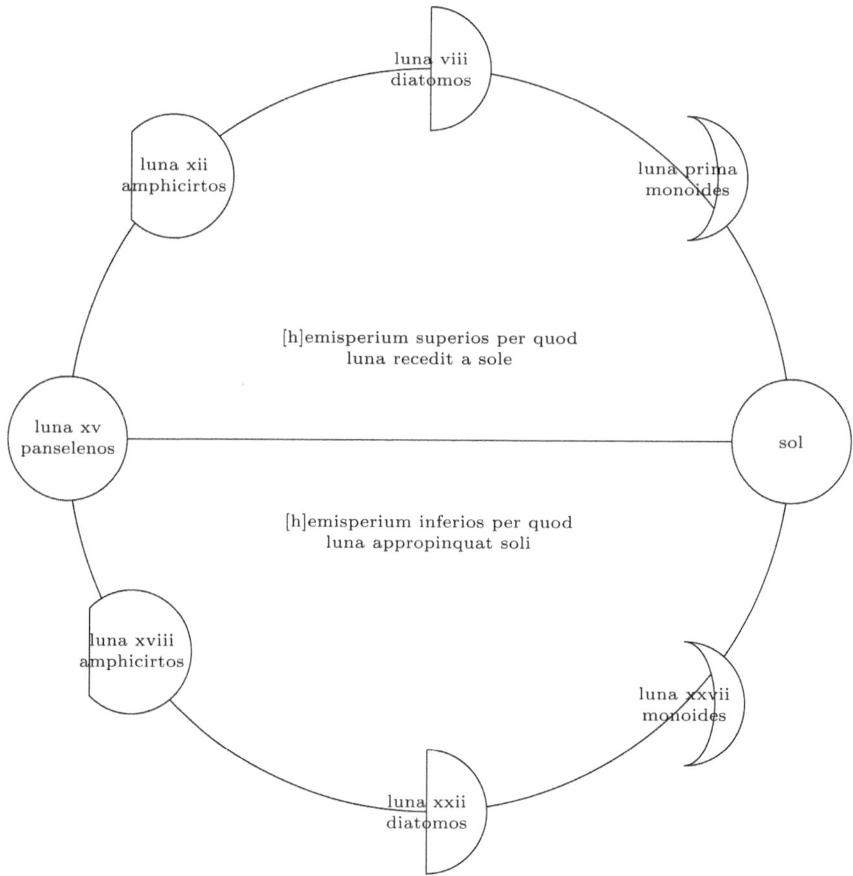

Figure 5.3 Re-elaboration of Abbo's diagram from Abbo, *Sententia de ratione spere*, ed. Thomson, 128.

Abbo's concise exploration of the Moon's phases also serves as an illustration of how he draws upon the various disciplines within the quadrivium to elucidate fundamental concepts in his philosophical system, like composition.

5.2 Artificial Compounds

In addition to examining natural compounds, Abbo also delves into what he terms artificial compounds. In Abbo's view, these compounds depend on human will (*secundum placitum*) and stick to the principles of an art or discipline – in this sense, one can call them 'artificial'. The expression *secundum placitum*, used by Abbo to describe this category of compounds, finds its origins in the writings of Boethius. In the second edition of his *Commentary on Aristotle's Peri hermeneias*, Boethius posits that a name does not carry meaning by nature, that is, inherently; rather, someone must impose a name to something to imbue it with significance, thereby creating an artificial and conventional sign.[22] Boethius also emphasises that determining the meaning of a name is not an arbitrary task undertaken by an individual. Instead, the process of naming necessitates recognition and acceptance of the name by a broader group of individuals.[23]

Abbo specifically attributes this conventional characteristic to the products of the quadrivial arts, that is, mathematical disciplines. Consequently, he exemplifies the category of artificial compounds with arithmetical inequality ratios, more particularly the five kinds of inequality ratios presented by Boethius in the *De arithmetica*. Abbo draws attention to these ratios to illustrate the artificial and conventional nature of compounds emerging from the realm of mathematics. This point becomes relevant in illustrating that the object of a discipline, particularly mathematics, seamlessly integrates into the broader context of nature and aligns with its inherent rationality. In this sense, Abbo states that artificial compounds resulting from human intellectual endeavours guided by the principles of a discipline, are in compliance with the nature because all discipline is, at its core, an imitation of nature.

The idea that knowledge resulting from a discipline complies with nature is influenced by another Boethian principle expounded in the *Commentary on Cicero's Topics*. There, Boethius argues that 'art imitates nature', emphasising that the art of topics, that is, of finding arguments ultimately relies on human rationality, which is clearly inherent in human nature. Consequently, not only the art of topics, but every form of reasoning following the principles of a discipline derives its ultimate subject matter from nature itself.[24]

Abbo extends this Boethian conception to articulate his view on artificial compounds. These compounds, originating from human will, align with nature insofar as they are crafted following the established schemes and rules of a discipline, which, in turn, depend on human nature.

> But I say that purposive things (*secundum placitum*) are those that are done or have been completed by will. These, too, do not differ from nature if they are reasonably constructed by art; since every art imitates nature, arising from the passions of the soul, which we believe to be natural. Hence it is from three equal terms that all species of inequality [ratios] are produced, and in the same order of their production, they are resolved into those very terms. For after equality, there occurs the multiple, namely, first species of inequality [ratio], second the superparticular, third the superpartiens, fourth the multiple superparticular, and fifth the multiple superpartiens – according to a three-step rule.[25]

Arithmetical inequality ratios, thus, are a kind of artificial compounds. Their consistency with nature and its inner rationality is evident from their composite nature: they can be produced and broken down following a precise procedure, adhering to certain arithmetical rules. As elaborated below, there exist distinct rules for their production and breakdown, enabling them to undergo a dual-phase development. They can become more complex through the application of a production rule and, conversely, be streamlined back to a simpler state – reducing them to the equality ratio (1:1) in the case of arithmetical products – through a reduction rule. This dual-phase development, involving a sort of expansion and contraction, is precisely what ties the natural and artificial compounds. Before delving into the specifics of producing and reducing arithmetical ratios it is prudent to elucidate certain fundamental arithmetical concepts.

5.2.1 The Five Kinds of Inequality Ratios

As it has been said, artificial compounds are illustrated by five types of inequality ratios involving two integer numbers. Abbo elucidates these concepts based on Boethius' *De arithmetica*, particularly the section focusing on 'related quantity'. In this context, related quantity consists in two numbers compared or linked to each other, so they constitute a ratio.[26] Boethius classifies related quantity into equal and unequal. Equal quantity arises when comparing two identical numbers, while unequal quantity results from comparing two different numbers that can be either greater or smaller than each other. Depending on the relationship between the greater and smaller numbers, five distinct kinds of arithmetical ratios emerge – commonly referred to as the five types of inequality ratios. These are: multiple, superparticular, superpartiens, multiple superparticular, and multiple superpartiens. These terms denote the greater number within the ratio, whereas the smaller number within the ratio is denoted by the same terms with the prefix *sub-*. In contemporary arithmetic, only the multiple and submultiple classifications have been retained. As per Abbo's definition, the multiple 'contains' the submultiple n-times, and conversely, the submultiple occurs n-times within the multiple. Clearly enough, multiplying the submultiple a specific amount of times results in a value that does not exceed or is less than its multiple.[27] Below, the definitions for the other four types of inequality ratios are provided: superparticular, superpartiens, multiple superparticular, and multiple superpartiens.[28]

> Given two natural numbers, a and b, where a is greater than b, the following ratios can be defined. In these definitions the aliquot $^{b}/_{n}$ is to be considered as an integer and, therefore, n is necessarily a factor of b.

- a is superparticular with respect of b when it 'contains' the latter plus one its aliquot part ($a = b + {}^{b}/_{n}$).[29] If $n = 2$, a is said 'sesquialter' and b 'subsesquialter'; if $n = 3$, a is called 'sesqui-third' and b 'subsesqui-third'; if $n = 4$, a is called 'sesqui-fourth' and b 'subsesqui-fourth'; and so on.
- a is 'superpartient' with respect of b when it 'contains' the latter plus x-times one its aliquot part ($a = b + c\,{}^{b}/_{n}$).[30] If $c = 2$ and $n = 3$, a is called 'super-two-third' and b 'subsuper-two-third'; if $c = 3$ and $n = 4$, a is called 'super-three-fourth' and b 'subsuper-three-fourth'; and so on.

- a is multiple superparticular with respect of b when it 'contains' y-times the latter plus only once one its aliquot part ($a = db + {}^{b}/_{n}$).[31] If $d = 2$ and $n = 2$, a is called 'double sesquialter'; if $d = 3$ and $n = 2$, a is called 'triple sesquialter'; if $d = 3$ and $n = 3$, a is called 'triple sesqui-third'; and so on.
- a is multiple superpartient with respect to b when it 'contains' y-times the latter plus x-times one its aliquot part ($a = db + c\,{}^{b}/_{n}$).[32] If $d = 2$, $c = 2$, and $n = 3$, a is called 'double super-two-third'; is $d = 2$, $c = 3$, and $n = 4$, a is called 'double super-three-fourth'; is $d = 3$, $c = 2$, and $n = 3$, a is called 'triple super-two-third'; and so on.

These ratios also formed the foundation of medieval music theory and were extensively discussed by Boethius in his *De institutione musica*.[33] Furthermore, the mathematical chess game known as *rhythmomachia* served as an instructional tool for acquiring a deep understanding of these arithmetical concepts.[34]

5.2.2 *Two* trinae regulae *for Arithmetical Ratios*

In the *De arithmetica*, Boethius delineates two procedures or rules by which all five types of arithmetical inequality ratios can both be derived from and reduced to the equality ratio. Each type of inequality ratio that makes the object of derivation and reduction can be represented by three numbers in the following way: the ratio between the first (x) and the second number (y) is the same as the ratio between the second (y) and the third number (z), so that $x: y = y: z$. Before delving into the explanation of the three-step rule (*trina regula*) of production and reduction of inequality ratios, let us first look at how Boethius elucidates how more complex inequality ratios stem from the equality ratio and from simpler inequality ratios.

> Let there be put down for us three equal terms, that is three 'ones', or three 'twos', or three 'threes' however many you want to put down. Whatever happens in one happens in the others. From these numbers. Then, according to the order of our precept, you will see born first the multiple and in these the duple first, then the triple, then the quadruple and so on according to that order. Again, if the multiple numbers are reversed, from these the superparticulars emerge, and from the doubles come the sesquialters, from the triples come the sesqui-third, from the quadruples come the sesqui-fourth, and the rest according to this manner. From the superparticular inverted it is necessary that the superpartients are born, and from the sesqui-fourth the super-four-partient proceeds. With the prior superparticulars placed rightly and not converted, the multiple superparticulars arise; with the superpartients rightly placed, the multiple superpartients are produced.[35]

Boethius' explanation can be restated as follows. If we consider a set of three equal numbers, these numbers express an equality ratio. Following a specific rule, from these three equal numbers, it is possible to obtain another set of three numbers expressing a multiple ratio: first the doubles, then the triples, then the quadruples, and so forth. If the sequence of these three multiple numbers is reversed and the production rule applied once more, three numbers expressing a superparticular ratio will result: from the doubles, one obtains the sesquialter, from the triples the sesqui-third, from the quadruple the sesqui-fourth, and so on. Once the order of the three superparticular numbers, too, is reversed, and the rule applied once again, three numbers expressing the superpartient ratio are produced: from the sesquialters the super-two-partient, from the sesqui-third the super-three-partient, the from the sesqui-fourth the super-four-partient, and so forth. From the superparticular and superpartient numbers in unchanged order, respectively, the multiple superparticulars and the multiple superpartients can be produced. Figure 5.4 aims to clarify the relations of dependence between these inequality ratios: the rightward arrow indicates that, within the derivation process, numbers are placed in a regular order, whereas the downward arrow indicates that the derivation occurs from numbers placed in reverse order.

As Figure 5.4 shows, from equal numbers only multiples can be derived. When three numbers are arranged in ascending order, multiples generate multiples, superparticulars generate multiple superparticulars, and superpartients generate multiple superpartients. Conversely, when the three numbers are arranged in a descending order, multiples generate superparticulars, and superparticulars generate superpartients. In other words, when we alter the order of the tern of numbers expressing the inequality ratio, by reversing it, then we obtain a tern expressing a more complex inequality ratio. If we do not alter the order of the tern of numbers expressing the inequality ratio, we obtain a tern expressing the corresponding multiple of the same kind of inequality. The production of inequality

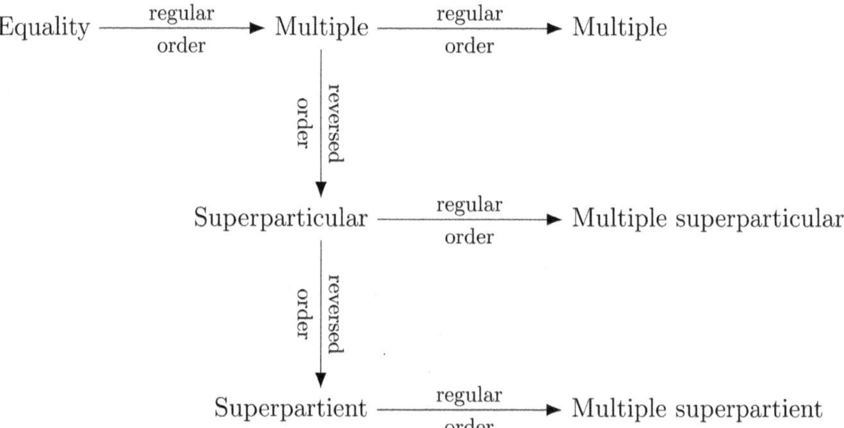

Figure 5.4 Derivation of inequality ratios.

ratios also considers the 'subgenus', i.e., the type of multiple from which they originate. For example, from the double only the sesquialter can be produced, and from the latter only the super-two-partient. Similarly, from the triple only the sesqui-third is produced, and from the latter only the super-three-partient.[36]

The Boethian rule for deriving inequality ratios from the equality ratio, that is, what we can call the production rule (P-rule), is taken up by Abbo in his *Explanatio* of Victorius' *Calculus*. It is called *trina* by Abbo, since it consists of three steps, which I will indicate respectively with P1, P2, and P3. Let us call the three given numbers a, b, and c; and the three derived numbers a_n, b_n, and c_n.

[P-rule]

Let us posit a set of three numbers a, b, and c, for example 2, 2, 2.

[P1] The first derived number (a_n) is equal to the first number of the original set of numbers ($a_n = a$); therefore, a_n is 2.

[P2] The second derived number (b_n) results from the sum of the first two numbers of the original set ($b_n = a + b$); therefore b_n is 4.

[P3] The third derived number (c_n) results from the sum of the first number (a), the second number multiplied by two ($2b$), and the third number of the original set (c); so that $c_n = a + 2b + c$); in this case c_n is 8.

In this way, from three numbers expressing an equality ratio (2, 2, and 2) we have obtained three numbers expressing an inequality ratio, more precisely, the double (2, 4, 8).[37]

By repeatedly applying this three-step rule, further inequality ratios can be produced. Figure 5.5 is introduced in the *Explanatio* as an example of this production rule. The table shows how all five kinds of inequality ratios (up to the double super-two-third ratio, namely the *duplus superbitertius*) are produced from three equal numbers.[38]

The table works in the following way. We start by placing three numbers expressing equality: 2, 2, 2. Following the production rule, we obtain three numbers expressing the double ratio in the second raw: according to [P1], 2 = 2; according to [P2], 2 + 2 = 4; and finally, according to [P3], 2 + 2(2) + 2 = 8. The second row is therefore formed by 2, 4, 8, which express, as already

2	2	2
2	4	8
8	12	18
18	30	50
8	20	50
18	48	128

Figure 5.5 Table based on *In Calculo*, 85.

mentioned, the double ratio. To derive the sesquialter (that is, the specific superpartient originating from the double), we then need to reverse the order of the doubles just obtained: 8, 4, 2. It is worth specifying that, in Abbo's chart, the steps concerning the inversion of the order of numbers are omitted – and this may make things a bit confusing. To the inverted doubles, the same three-steps rule [P1-3] is applied, so that we obtain 8, 12, 18 for the third row. To obtain the three number expressing the super-two-third ratio (*superbitertius*, that is, the specific superpartient ratio originating from the sesquialter), we need to reverse the order of the sesquialter themselves (18, 12, 8) on the third row. Applying the P-rule again, we obtain the super-two-thirds in the fourth row: 18, 30, 50. Now, to get the last two inequality ratios, we need to go back to the second and third rows – this, too, is omitted in the table. Without changing the order of the sesquialter and the super-two-thirds numbers, the multiple superparticular (18, 30, 50, expressing the double sesquialter) and the multiple superpartient (18, 48, 128, expressing the double super-two-third) are respectively produced by applying the production rule [P1-3].

The latter species, i.e., the double super-two-third, is a 'subgenus' of the multiple superpartient, and in Abbo's table represents the highest level of complexity among inequality ratios. According to Abbo, once this level of complexity is attained, artificial compounds exhibit behaviour similar to natural compounds, to the extent that they undergo a process of decrease, ultimately returning to the equality ratio through a reduction rule. From this perspective, the multiple superparticulars can be likened to the peak of human strength at the age of 35 and to the full moon. This marks a turning point in Abbo's argumentation: the various inequality ratios, emblematic of artificial compounds, diminish akin to natural processes, retracing the same developmental stages in reverse. Another arithmetical rule, denoted as R-rule, serves as the inverse counterpart to the P-rule. While the P-rule transforms equality ratios into inequality, the R-rule facilitates the reduction of inequality ratios back to equality. Therefore, the process of reducing arithmetical ratios parallels the waning of human strength and the gradual dimming of the moon.

The R-rule, too, consists of three steps (R1, R2, and R3). Given a set of three numbers expressing an inequality ratio (a, b, and c), I denote a_n, b_n, c_n as the three 'reduced' numbers expressing a simpler ratio.

[R-rule]

Let us posit a set of three numbers (a, b, and c), such as 18, 48, 128, which express an inequality ratio, such as the double super-two-third ratio (*duplus superbitertius*).

[R1] The first reduced number (a_n) is equal to the first number of the original set of three ($a_n = a$); thus a_n is 18;

[R2] The second reduced number (b_n) results by the subtraction of the first term of the original set of three (a) from the second of the same set (b), so that $b_n = b - a$; therefore, b_n is 30.

[R3] The third reduced number (c_n) results from the subtraction of the first number of the original set of three (a) and twice the second reduced number ($2b_n$) from the third number of the original set of three (c), so that $c_n = c - a - 2b_n$; therefore c_n is 50.

In this way we have reduced the double super-two-third (*duplus superbitertius*) to the super-two-third (*superbitertius*), going from the multiple superpartient ratio to the superpartient.[39]

The application of the reduction rule (R-rule) as explained by Boethius posed a significant challenge for Abbo and some of his contemporaries, including Gerbert of Aurillac and Notker of Liège (c. 940 – 1008). This challenge arose from an interpretive lacuna in Boethius' *De arithmetica.* There, Boethius mentions the reduction rule, stating that "if one takes a superparticular sesqui-fourth (*superparticularis sesquiquartus*), first they turn back to the sesqu-third (*sesquitertius*), then to the sesquialter, and finally to three equal terms".[40] However, Boethius provides an example only for the case where quadruples are reduced to triples, then to doubles, and finally to equal terms, excluding superparticulars and superpartient.[41] This omission is attributed to a fundamental limitation in the R-rule: it does not address the reduction of the four more complex inequality ratios to equality but only applies to multiples reduced to equality. Boethius, in formulating the R-rule, fails to specify the precise way types of inequality ratios more complex than multiples could be reduced to equal numbers.

Despite this limitation in Boethius' explanation, Abbo's exploration of the R-rule entails a demonstration of the reduction process for three numbers expressing the double super-two-third ratio (18, 48, 128, *duplus superbitertius*, which is a superpartient) to three numbers expressing the super-two-third ratio (18, 30, 50, *superbitertius*, which is a simpler type of superpartient). Then, he vaguely refers to the subsequent derivation of the superparticular, the multiple, and finally equality from these latter numbers.[42] While presenting the problem using Boethius' words from the *De arithmetica*, Abbo specifies that, although Boethius does not acknowledge any intermediate steps between a sesqui-fourth (*sesquiquartus*) and a sesqui-third (*sesquitertius*) to reach back to equality, this does not preclude the existence of intermediate steps involving respective multiples to jump from superpartients to superparticulars.[43] A full explanation, however, is omitted. This vagueness, as much as posed by Boethius' ambiguity, merely postpones the challenge to Abbo's conceptualisation of compounds. In the absence of an effective way to reduce all inequality ratios to equality, the risk of a lack of a precise correspondence between artificial and natural compounds remains evident.

Abbo's survey of this aspect of Boethian speculative arithmetic did not stand alone in scholarly interest. Two contemporaries, Gerbert of Aurillac and Notker of Liège, also delved into this subject.[44] Gerbert presented his method of reduction, known as the *saltus Gerberti*, in a letter addressed to Constantine of Fleury.[45] More precisely, Gerbert's R-rule involved alternating steps between the P-rule and the R-rule. The procedure commenced by reducing the sesqui-fourth to the quadruple (R-rule), subsequently reduced to the triple (R-rule). From the triple, Gerbert produced the sesqui-third (P-rule) and obtained the double (R-rule). The double was then reduced to equality (R-rule), and, to encompass all inequality ratios, the sesquialter was produced (P-rule). Having hitherto applied the P-rule and R-rule alternately, Gerbert is able to reduce the superparticular, that is, the sesqui-fourth to the quadruples, the triples, the doubles, and finally the equality. By consistently applying the P-rule and R-rule alternately, Gerbert successfully reduced the superparticular, specifically the sesqui-fourth, to quadruples, triples, doubles, and ultimately equality. However, Gerbert's approach differed from Boethius' assertion, as it exclusively descended from the sesqui-fourth to equality via other simpler superparticulars. In his work *De superparticularibus*, Notker, a Benedictine monk from the St. Gallen monastery who became the Bishop of Liège in 972, formulated a method more intricate than Gerbert's. Notker sought to adhere closely to Boethius' teachings in *De arithmetica* by deriving simpler superparticulars from the sesqui-fourth without 'passing through' multiples. In doing so, though, Notker introduced a new R-rule, deviating from the Boethian framework.[46]

Gerbert's and Notker's discussions of the Boethian R-rule are strictly mathematical, as they are not embedded in broader philosophical considerations relating to any arithmetical and natural compounds – to use Abbo's vocabulary. This appears in fact to be a distinctive trait of Abbo's philosophical system. Other early medieval commentaries on Boethius' *De arithmetica*, such as the *Ne subesse possit* (linked to Eriugena's circle by Irene Caiazzo), the *De initio* commentary (linked to the milieu of Gottschalk of Orbais by Jeremy Thompson), the 'Florentine' commentary (preserved in Florence, Bibl. Laurenziana, Plut. 51, 14 and sharing strong similarities with early medieval commentaries on Boethius' *Consolation*), and Thierry of Chartres' commentary – none of them addresses the P- and R-rules focusing on the strictly mathematical content, nor does any of them set forth a parallelism between such arithmetical rules and natural phaenomena. In this respect, it is Abbo's originality, then, to intertwine speculative arithmetic and philosophical stances on nature. As it is his originality to turn to arithmetical rules to elucidate what composition is – a concept mobilising other philosophical concerns, such as the ontological discrimination between *esse* and *id quod est* that is analysed in the next section.

5.3 Ontological Composition: The Case of the Earth in *Genesis* 1:2

The dynamics of growth and decrease, generation and corruption of natural compounds are mirrored by the production and reduction of arithmetical ratios. This parallelism allows Abbo to emphasise another essential characteristic of all compounds: each of them has an origin to which they can all be traced back. Abbo articulates this point through a categorical syllogism, presented below.

> Every compound is joined by its parts. However, the joining of parts anticipates dissolution either by nature or by will. But what can be broken down by nature is known to return to its origin. Therefore, every compound by nature is dissolved into its primordial state. There is also something that can be broken down by will, which exists in imitation of nature. Moreover, the natural dissolution does not depart from the principle of its own connection. For all mutable things reach their peak and return to where they began. Therefore, by imitating this, what can be broken down [by will] is bound to the origin of its composition. For no compound lacks an origin. Also, everything that can be broken down is a compound. Thus, nothing that can be broken down lacks an origin. Hence, it is evident that nothing lacking an origin can be broken down.[47]

Looking at Abbo's syllogistic reasoning, it becomes apparent that every compound arises from the union of its parts and thus, is dissolvable (*solubilis*) in the sense that it can be broken down into its constituent parts. The process of dissolution precisely involves disassembling the compound into its fundamental components, ultimately reverting it to an original state that precedes the assembly and connection of its parts. The dissolubility by will (or artificial dissolubility) imitates the natural dissolubility, for all compounds have the potential to revert to their original state or origin. Consequently, entities lacking an origin, as they lack further dissolvable components, cannot be deemed compounds. Anything that does not have any origin, thus, is not composite.

The process of dissolving (*resolutio*) consists in the breaking down of a compound into its fundamental constituents, so that the compound returns to its original state that precedes its composition. It is interesting to notice that Abbo's distinctive philosophical notion of the breakdown of compounds might have served as the philosophical backdrop for an anonymous glossator of Macrobius' *Commentary on the Dream of Scipio*. In the 11th-century manuscript Paris, BnF, lat. 7299, a marginal note on fol. 32r reads: "For of all things anything that can be dissolved is by necessity composite. But what is composite is not simple. Therefore, simple and absolutely non-dissolvable is that which is not subject to any division nor admits any composition".[48] This gloss interprets a passage from Macrobius' *Commentary* (I, 5, 17, more precisely the word *solvitur*) concerning the breaking down of the number eight into equal parts, implying its potential repeated halving to achieve simple unity (eight being an 'even times even' number per Boethius' definition).[49] The glossator does not delve into the arithmetical implications of the text but rather focuses on the expression 'to be dissolved' (*solvitur*), introducing the contrasting notions of simplicity and composition. They posits that anything capable of being dissolved is inherently composed, while simplicity precludes any division. Now, scholars like Lapidge and Barker-Benfield suggested Ramsey and Fleury as potential origins for the Parisian manuscript.[50] Considering this insight, the gloss can be regarded as evidence of the impact of Abbo's distinctive philosophical theories on the interpretation of Macrobius' *Commentary*. As we are going to see, Abbo's conceptualisation of composition, wherein compounds are viewed as dissolvable into constituent parts, not only underpinned the interpretation of philosophical works like Macrobius', but also extended to some passages of the Bible, introducing what one might call the 'ontological' composition that holds for all created things.

As an illustration of the division into parts concerning compounds, Abbo highlights instances where material entities, though seemingly intact and solid, are, in fact, compounds. Examples of such material entities include the 'earth' or a 'stone', which are compounds both in a physical and an ontological sense. Physically, they consist of heterogeneous components whose conjunction gives rise to a physically cohesive entity. Ontologically, their structure involves 'being' (*esse*) and 'that which is' (*id quod est*). Abbo delves into this matter drawing from *Genesis* 1:2, interpreting it in a distinctive manner influenced by Boethius' *De hebdomadibus* and *Commentary on the Categories*. The 'earth', considered as a composed physical element, takes centre stage in this discussion.

In the Septuagint version of *Genesis*, as cited by Abbo, it is stated that "the earth was invisible and incomposite".[51] Abbo interprets this passage consistently with his views on the composition: earth cannot be incomposite in an absolute sense, for it indeed has an origin. Thus, he posits that the Biblical account refers to the earth as a physical element, specifically in its primordial state. In Abbo's understanding, the earth was initially intermingled with other elements in a confusing fashion, rendering it indistinguishable from them. Consequently, the earth was described as 'invisible and incomposite', as reflected in the *Genesis* narrative. According to Abbo, the earth's lack of composition stems from its undifferentiated state with the other elements. The earth's capacity of being visually perceptible emerges only with differentiation brought about by its specific qualities, namely, coldness and dryness.

> But we read "the earth was invisible and incomposite". Which is undoubtedly this way, if [the earth], before its separation from the other elements, was incomposite and invisible due to some indiscriminate mixture. Once visible, it began to be composite, not indeed by other elements but by its qualities.[52]

In his interpretation of *Gen.* 1:2, Abbo considers the earth as one of the four elements. This interpretation stands out as original, particularly when compared to the preceding medieval exegetical traditions. Let us briefly revisit some of the accounts of the biblical earth that Abbo might have been familiar with. In the *De Genesi ad litteram* and the Book XII of the *Confessions*, Augustine interprets the earth mentioned in *Gen.* 1:2 as formless prime matter. According to Augustine, this earth corresponds to a capacity to receive forms. While preserving a minimal ontological capacity, matter is informed by God through the eternal rational principles existing in his mind.[53]

Unlike Augustine, in the doxographical section of his *Commentary on Plato's Timaeus*, Calcidius directly refers to the *Genesis* and presents at least four ways of conceiving matter: the term 'earth' in the Bible could stand for the sensible earth observable in the universe, a sort of intelligible paradigm, the substance of bodies (referred to as "silva corporea") lacking form and quality, or intelligible matter.[54]

A different perspective is offered by Bede. In his *Commentary on Genesis*, Bede contends that the term 'earth' in the Vulgate version of the Bible, described as "inanis et vacua", refers to the whole material reality, that is, the world still in a state of emptiness. In its primordial state earth was entirely covered by water, and both elements – earth and water – shrouded in darkness. Earth and water were not the only two elements to come first into existence simultaneously, for Bede posits that fire and air were there as well. However, Bede does not hold that these four elements in their primordial state were mixed and formless: for instance, earth, confined on all sides, already appeared as it is now at the bottom of the sea. Earth, like water, could be considered formless matter only to the extent that form and beauty of all created reality are brought forth only when light is created.[55] Bede's insights are echoed almost verbatim in the commentaries on *Genesis* by Remigius of Auxerre and Hrabanus Maurus.[56]

Another distinct reading of *Gen.* 1:2 is presented by Eriugena in the *Periphyseon*. Eriugena approaches this biblical verse figuratively (*figurative*), viewing the earth as one of the primordial causes of visible things. The description of the earth as empty signifies the utmost subtlety and simplicity of its intellectual nature.[57] In his survey of *Gen.* 1:2, Eriugena also covers the perspectives of other thinkers, including the interpretation he attributes to Basil of Caesarea. According to this view, *Gen.* 1:2 expresses the state of the world's body before it was adorned with the various shoots, fruits, and animals of diverse kinds and species.[58]

The various interpretations of the Biblical earth recalled so far do not align with the 'physical' reading proposed by Abbo, nor do they encompass his 'metaphysical' interpretation, which stands as his distinctive contribution and will be now discussed. As anticipated, Abbo connects his exegesis of *Gen.* 1:2 with his interpretation of Boethius' *De hebdomadibus* and *Commentary on the Categories*.[59] Through a Boethian lens, Abbo presents his perspective on the cause of the existence of creatures, their composition, and their qualitative determination. According to Abbo, both visible and invisible created things owe their existence to an immaterial cause: God. Existing independently from everything else, God is simple, as in him, 'being' (*esse*) and 'that which is' (*id quod est*) coincide. In contrast, creatures cannot be considered simple, as they receive the 'form of being' (*forma essendi*), resulting in the separation of their 'being' and 'that which is' into two distinct components. Abbo further develops this determination of composition by introducing what might be interpreted as a distinction between the existence and essence inherent in every creature. He articulates this point as follows: on the one hand, creatures exist materially as concrete entities ("materialiter per rem"), and on the other hand, they are essentially specified through substantial quality ("essentialiter per substantialem qualitatem").

> For while there is a certain cause of all visible and invisible things, which by itself is being by which it is what it is, it never lacks in creatures even if it is not material, combining each thing with its forms, by which things can be defined. Thus, it happens that, since created things do not subsist by themselves, after receiving the form of being, they are not simple. For in the case of this stone, its being is one thing, and what it is, is another; for one pertains to its material nature as a thing, and another pertains to its essential nature as a substantial quality. Thus, true is that claim which – by aetiology, namely, by an explanation of the cause – suggests that everything is necessarily subject to division because it is a compound, namely, composed of various constituents, as has been said. For 'being' is different from 'what it is'. This composition also provides order to things, without which all things would be vain, empty, and formless.[60]

Abbo contends that this composition of 'being' and 'that which is' is responsible for imparting an orderly beauty (*ornatus*) to created things. According to him, the structured beauty of the created reality ultimately derives from this ontological composition, allowing each created thing to exist in harmony with its 'that which is' (or essence, one could say), thereby rendering it beautiful. A similar emphasis on the ontological composition resulting from the reception of the form of being (*forma essendi*) is evident in the glosses to Boethius' *De hebdomadibus* attributed to Eriugena. According to these Carolingian glosses, all things exist in the divine mind before being created and specified by the form of being. Despite this, God is associated with 'being' (*esse*), being absolutely simple. In contrast, all composition of parts is linked to 'that which is' (*id quod est*), and this composition is evident from the moment creatures come into existence, namely, when each creature is spatially and temporally determined.[61] Similar to the Carolingian glossator, thus, Abbo asserts that creation itself implies a composition for things due to the reception of the form of being, resulting in their specified essence.

With these observations, Abbo seeks to demonstrate that everything in material reality is composed and, consequently, can be broken down into its distinct parts or constituents, both in a material and an ontological sense. Notably, he adeptly combines Biblical exegesis with Boethian ontology to construct his own perspective on reality, characterised primarily by its inherent composition. It is worth noting that this feature – that is, being composite – extends not only to entities existing by nature, but to those existing by human will, as discussed below.

5.4 Division of Abstract Entities: The Case of the Units of Time

Composition is intrinsic to everything within created reality. If unity can be somewhat attributed to entities other than God, this is only on account of their being a whole, that is, on account of their being a unity which is still fractionable. Whether an entity is a physical quantity or a unit of measurement of weight or time, it remains subject to fractioning or breakdown into constituent parts. Compounds can be both entities that exist naturally ("naturae beneficium") as well as abstract ones shaped by human reason ("rationis patrocinium"). Abbo delves into the examination of this second type of entities – let us call them abstract – that are still defined by quantity. In doing so, he relies on two disciplines, namely, logic and astronomy, using the day (*dies*) and the hour (*hora*) as case studies.[62]

Abbo's discussion of divisible abstract entities commences with the assertion that these entities lack corporality and cannot be considered substances properly speaking. Regarding the case-study of the day (*dies*), he initially adopts Bede's definition from *De temporum ratione*, wherein the day is described as 'the air illuminated by the light of the sun', with its opposite being the night.[63] Abbo employs Boethius' *Commentary on Aristotle's Categories* to clarify that terms in an oppositional relationship mutually exclude one another – meaning the absence of one term determines the presence of the other. Substances, like humans or air, can, at different times, exhibit opposites such as health and illness or day and night.[64] Substances themselves do not possess inherent opposites. Therefore, the day is not a substance but a quality of light intrinsically associated with the sun, illuminating the air when passing over the earth.[65]

While not constituting a substance, Abbo argues that what we call 'day' can still be subject to division. Drawing on Cicero's definition of time from his *De inventione*, Abbo posits that 'time' is a term articulated in accordance with a conventional meaning contingent upon specific durations, such as the length of the day, night, or year.[66] Within this framework, hours serve as internal components of these durations and can be quantified through a practical proof or test (*experimentum*), involving the use of a water-clock. Abbo attributes the origin of this practical proof to unspecified 'ancestors' knowledgeable in astrology. It is plausible that the ground for such a reference is the astronomical section of Macrobius' *Commentary on the Dream of Scipio*, where the method employed by the Egyptians for dividing the Zodiac into 12 signs is discussed.[67] The method described by Macrobius aligns in fact with Abbo's exposition in his *Explanatio* and entails precisely the utilisation of a water-clock. It is perhaps no coincidence that in two manuscripts that can be associated with Fleury and transmit Macrobius' work, one and the same gloss was copied in the margin of the passage of Macrobius' work that presents the Egyptian method (I, 21, 12): manuscript Paris, BnF, lat. 7299 (f. 51r) – which preserves the marginal gloss on the Abbonian meaning of dissolution mentioned in the previous section – and manuscript Paris, BnF, lat. 16677 (f. 38v).[68] Notably, this gloss defines a water-clock in line with Abbo's description, namely, as a vessel from which water flows out through a hole: "clepsidra est vas unde quasi furando deorsum trahitur aqua". Considering Abbo's technical explanation of the division of the day – which is addressed below –, glosses of this type are further evidence that Macrobius' *Commentary* was read at Fleury and that Abbo's view on composition oriented the interpretation of such passages in Macrobius' text.

The water-clock is an instrument employed since antiquity to measure time, combining the observation of the flow of a liquid from a container with the monitoring of celestial movements.[69] Let us see for what purpose and in what way this instrument was used according to the description offered by Abbo. The water-clock was utilised to divide the equinoctial day and night – representing the entire revolution of the celestial sphere on the spring and autumn equinoxes – into 24 equal parts, each termed 'hour' (*hora*). Moreover, 'hours' could in turn be subdivided into 'points' (*puncti*) and 'parts' (*partes*). This technical explication of the fractioning of hours into points and parts can be considered against the backdrop of Bede's *De temporum ratione*. Bede explains that within each solar hour (60 minutes) there are four points (1 *punctus* = 15 minutes), and each point contains three parts (1 *pars* = 5 minutes). However, in the context of the lunar cycle, an hour could be divided into five points.[70]

> 'Hour' marks the end of a portion of time, which, although it is a certain part of eternity, is nevertheless expressed according to a specific meaning of a segment of it, such as the span of the year or of the day. For when our ancestors, devoted to astrology, divided the equinoctial day and night together into twenty-four equal parts using a water-clock, they called the end of each part 'hour', with aspirated 'h'. They also recognized that the entire mechanism of the heavens revolves daily through these 24 parts or hours. They called water-clock [*clepsidra*] any bronze vessel, through whose base, a very small opening was made, allowing water to flow out as though it were being stolen – they interpreted *clepere* as to steal and *ydor* as water. By the careful collection and division of that flow of water, as Macrobius attests, they demonstrated with a practical experiment the course and return of the entire celestial sphere, along which the oblique order of the zodiac signs has always turned, assigning two hours to each of the twelve signs. For the sake of greater ease in calculation, they also divide the span of the hours into four or five 'points', and one point into three 'parts'.[71]

In a monastic setting, especially for fulfilling nocturnal religious duties, the primary method of measuring night-time was through the observation of stars.[72] Given the absence of specific tools for time measuring, monks could gauge time by noting the change in the azimuth of a star (i.e., the angle measuring a star's movement on a horizontal plane) from designated locations within the monastery ("locus designatus"). This practice is well-documented in a liturgical chants text from the abbey of Fleury known as the *Horologium stellare monasticum*, dating back to the 11th century.[73] Alongside this observational approach, Fleury is noted for employing the water-clock, as evidenced not only in Abbo's *Explanatio* but also in the *Consuetudines floriacenses antiquiores* by the monk Thierry, who discusses the water-clock in the context of the duties of the vigil monk.[74] The use of the water-clock is also attributed to monasteries other than Fleury's, such as the one of Reims. This is evidenced by a letter by Gerbert of Aurillac from 989 addressed to his pupil Adam at Reims, where Gerbert mentions *horologia* detailing the temporal duration of day and night hours in two locations with distinct geographical latitudes. Gerbert expresses the hope that Adam, with the assistance of the water-clock, will continue measuring the duration of day and night during solstices in various geographical areas (*climata*).[75]

In the *Explanatio*, Abbo provides a comprehensive description of the functioning of the water-clock and its application in timing the rising and setting of zodiacal stars, along with calculating the duration of an entire revolution of the celestial sphere.[76] The process is outlined as follows. When a specific star appeared on the horizon, a bronze vase was filled with water which was allowed to flow through a designated opening at the bottom of the vase. This water flow continued throughout the night and day until the reappearance of the same star, signalling a complete revolution of the sky. The water was collected and divided into 24 equal parts, eventually being poured back into the emptied vase. Subsequently, two smaller vessels, each holding one 24th of the collected water, were utilised. At the next rise of the designated star, the water flowed from the larger vase into the first small vessel. Upon filling, the first vessel was exchanged with the second, and this alternation was carried on until the water was depleted. Each time the second vessel was emptied, that indicated two hours has passed, marking the movement of a zodiacal sign.

> Such consideration was conducted in this manner: the aforementioned bronze vessel was filled with water while its opening was sealed. Once carefully prepared, and having first forecast a clear weather, one of the brightest fixed stars was selected. Therefore, at the rising and appearance of this star, the opening of the vessel was unsealed, and the water immediately began to flow out. This continuous flow of water lasted throughout the entire night and the following day until the same star rose and appeared again, marking the complete revolution of the entire

> celestial sphere. The water that flowed out was then carefully collected so that, evenly distributed into twenty-four parts, it could be poured back into the same emptied bronze vessel. After three or however many days, taken two vessels, each capable of holding a twenty-fourth part of that water, the rising of the designated star was noted and immediately the flow of the water was started again. When one twenty-fourth part was collected in one of the two vessels, it indicated that one hour had elapsed. To ensure that not even a single drop of water was lost while transferring the water from one of the small vessels, the other was placed underneath so that, when it was full, it signified the passing of the second hour. During the span of two hours, without a doubt, one of the twelve signs of the Zodiac would rise or set. Thus, through the alternating use of vessels, both the duration of the hours and the size of the zodiacal signs were determined.[77]

In his *Commentary*, Macrobius mentions that the water was collected and distributed into 12 equal parts, indicating not the hour, as in Abbo's description, but two hours.[78] This variation stems from the distinct objectives of the two authors. Macrobius aims to illustrate how the Egyptians divided the Zodiac into 12 equal segments, while Abbo seeks to establish that an hour constitutes the 24th part of the day, quantifiable despite being an entity established not by nature, but by human will only. Abbo further expounds on the outcomes of this test, highlighting that the distribution of hours during the day and night varies based on the Sun's position along the ecliptic. For instance, during the spring and autumn equinoxes, both diurnal and nocturnal hours amount to 12. In contrast, at the summer solstice, there are 18 diurnal hours and 6 nocturnal hours, and this distribution reverses at the winter solstice.

> Indeed, at the time of the equinox, half of the total water would flow out completely during the night, and the other half completely during the day. At the summer solstice, six parts of that water would flow out in the night, and the remaining eighteen parts during the day. At the winter solstice, the opposite occurs, with eighteen parts flowing during the night and six remaining for the day. Furthermore, the increase or decrease of the length of each day was revealed by adding or subtracting some ounces or fractions of ounces [*minutiae*] in the water flow. Since things are so, it is certain that 'day' and 'hour' are terms referring to time, while time, like body, is a species of quantity – not in that it 'is' but in that it 'is a quantum'.[79]

As previously noted, Gerbert also addresses timekeeping in his letter to Adam. He presents his pupil with various *horologia* related to two locations situated at different latitudes: the Hellespont (Dardanelles) and another unspecified site where daytime hours extend to 18 on the longest day of the year. Explicitly rejecting the theory of those who disregard such distinctions and consequently believe that hours remain the same throughout each month of the year, Gerbert adheres to the theory of Martianus Capella. According to Martianus, the length of the day and night varies based on both the month (i.e., the position of the sun along the ecliptic) and the geographical location where the measurement is taken.[80] Quoting Martianus, Gerbert asserts that the duration of daylight, i.e., the time the sun is above the horizon, increases progressively starting from the winter solstice. According to this pattern, in the first subsequent month, it increases by a 12th of the difference between night and day hours, in the second month by a sixth, in the third and fourth month by a quarter, and in the fifth and sixth month by a sixth and a 12th.[81] As evident from the previous quote from the *Explanatio*, Abbo also anticipates the gradual extension and reduction of daytime hours. He specifies that the quantity of water flowing from the water-clock, corresponding to the daytime hours, can vary by ounces or *minutiae*, indicating duodecimal parts of the same quantity. In the light of this observation, Abbo cannot be categorised among those criticised by Gerbert for failing to recognise changes in the distribution of hours per day and night throughout the year. What is noteworthy is that both scholars draw similar conclusions about the varying length of the day and the distribution of daylight hours throughout the year, even though they rely on two distinct astronomical traditions – Abbo on that introduced by Macrobius and Gerbert on that developed by Martianus.

In concluding the discussion on timekeeping, Abbo asserts that the terms 'day' and 'hour' denote specific durations and can be regarded as divisions or fractions of time. Analogous to the nature of the body, time, according to Abbo's interpretation of Boethius' *Commentary on the Categories*, can be understood as a species of quantity.[82] Abbo contends that time, emblematic of non-corporeal entities, qualifies as a quantity precisely because it is divisible and can be thus expressed numerically. Hence, the concept of quantity becomes pivotal in determining whether a given entity, whether corporeal or not, exhibits a composite nature. The next section delves into the quantification of only apparently non-corporeal entities, such as the sound or voice.

5.5 Divisibility and Corporality: The Case of the *vox*

The category of quantity plays a pivotal role in discerning corporeal entities. According to Abbo, any natural thing that presents itself as a solid whole is such on account of the presence of all its parts and its thickness (*crassitudo*). Abbo extends this perspective to the minutest things in nature, exemplified by a cobweb thread – in this, he paraphrases Claudianus Mamertus, who in the *De statu animae*, emphasises that even extremely small entities exhibit the attributes of a body.[83] Consequently, in Abbo's view, perceptibility by sight alone is an insufficient criterion for categorising something as corporeal since not all bodies are perceptible by sight.

> We are used indeed to ascribe wholeness to something that we know has neither division nor cutting nor lessening. We also say that it has solidity, if we usually mean by it the thickness of a body, which nothing visible lacks, not even a spider web. This comes as no surprise, since every visible thing is corporeal, even though not everything corporeal is visible.[84]

The defining characteristics of a body, according to Abbo, are its perceptibility by the senses and quantifiability, expressed through divisibility. Anything that the soul apprehends through the body is considered a body, and this includes entities perceptible through sight. Thus, everything visible is necessarily corporeal. Furthermore, every body is subject to measurement and division, falling under the category of quantity. In contrast, entities that lack quantifiability and divisibility, like the soul, are deemed incorporeal. Abbo illustrates these principles using the example of the human voice: although invisible, the human voice is not incorporeal, as it is still perceived through the sense of hearing and, crucially, is subject to quantification, manifesting variations in pitch according to precise numerical criteria.

> Everything that the soul perceives through the body must necessarily be corporeal. Indeed, the soul employs the various functions of the body (as its servants) through which sensory things, i.e., things that can be perceived by external senses, are reported to it as an inner master. These are reported, for instance, in the case of all sensory objects that are visible. Therefore, every visible things is corporeal. [...] Hence, it must be inquired what things are incorporeal, which can be numerically divided. For nothing is subject to number except a body. [...] Moreover, what is not subject to number is in no way divided into parts. Something incorporeal is indeed not subject to number and thus cannot be divided into parts. But the invisible voice itself is made lower or higher pitched by numerical proportions, so that it may be pleasing to hear. If it can be heard, it is sensory. And everything sensory is a body. Hence, it is clear that the voice is a body.[85]

Quantity is key to identifying those corporeal entities, such as the human voice (*vox*), which lack visibility. When Abbo asserts that the voice is measurable and subject to number, he appears to highlight its modulability, indicating that its pitch can be varied systematically and expressed in a numerical series or proportion. This emphasis on numerical proportion suggests a connection to the musical scale, especially in the context of Abbo's time when it was predominantly applied to chanting. Therefore, Abbo's consideration of the voice as a corporeal entity subject to number may be specifically directed towards the human voice rather than sound in general. This interpretation is reinforced by a passage from the *Explanatio* where Abbo describes situations where the sense of sight seems to be confused with hearing – in cases when one exclaims “see how well they sing”. In such instances, the recognition of harmonies through hearing enables one to somewhat ‘see’ how the voice is arranged in accordance with a numerical order.[86] Being perceptible through hearing and characterised by numerical modulation, the voice can be regarded as a quantity and, consequently, as a bodily substance. The case of the voice underscores that certain invisible natural things are corporeal not only due to their perceptibility through senses other than sight but also because of the numerical structure defining them – that is, because of their being quantified, a quantum.

This tight link between corporality and quantity, exemplified by the case-study of the human voice, is also accounted for through logic. Abbo argues that since a substance can only acquire the accident of quantity through a body, ‘body’ and ‘quantity’ can be regarded as peculiar properties (*propria*) that are logically interchangeable within the definition of substance.[87] In the framework of Aristotelian and Porphyrian logic, the term *proprium* refers to one of the five predicables that define a substance (i.e., genus, species, peculiar property, difference, and accident). The *proprium*, thus, represents a specific logical predicate permanently attached – yet not necessarily or essentially – to all individuals of the same species. For instance, a peculiar property of human beings could be laughing or the ability to use grammar, which are features that are unique to this species.

Abbo conceives quantity and corporality as peculiar properties inherent to each substance. He asserts that where there is a body, there must also be quantity, emphasising a strong interconnection between ‘body’ and ‘quantity’. This becomes clear if one considers how the soul comprehends an object, distinguishing three aspects of it: that it exists (*esse*), that it is of a certain quantity (*esse quantum*), and that it is of a certain quality (*quale esse*). Abbo contends that the soul’s recognition of something as a body hinges not on its existence or quality but fundamentally on acknowledging its quantitative being. Quantity, as Abbo conceives it, is not limited to physical extension but encompasses a broader notion of a structure responding to some numerical rationality.

> Hence, it is clear that the voice is a body, although someone might object with reasoning of this sort: “Silence”, they say, “is opposed to voice, just as darkness is to light. However, no opposite is a substance. But the body is a substance, like a stone. For a stone is a substance. Therefore, the voice is not a body”. We briefly refute this objection, because spending effort in such matters is like wasting effort in looking for a knot in a bulrush. Every corporeal substance is so surrounded and included by those things

> it underlies that it cannot be perceived by sight without them [...]. Substance does not need quantity except through the body, for quantity and body are interchangeable as peculiar properties. For wherever there is quantity, a body cannot be absent, and wherever there is a body, there must necessarily be quantity. Therefore, it is known to all wise people that body is an attribute of an accident, not of a substance, because its being is properly quantity. And this difference is intelligibly grasped by the soul in one and the same thing, for example, a stone. For to the same stone, being, being quantified, and being qualified are different things.[88]

There is another counterargument that Abbo addresses. If one assumes that the voice is a quantified substance, they should also recognise that voice logically does not possess an opposite, given that neither the substance nor the accident of quantity inherently oppose anything. Yet it seems that silence is the opposite of voice. To unravel this theoretical intricacy, the following argument is presented, relying on Boethius' logical reflections, with a particular emphasis on the category of relation.[89] A substance or an accident such as quantity, considered in themselves, does not have opposites. However, when viewed in relation to something else, they can, to some extent, be positioned in opposition to something. For instance, Socrates himself does not inherently possess an opposite, yet when considered as a father, he becomes somewhat the opposite of his son – i.e., he is in relation to his son. Similarly, the number four, which is an instance of quantity, does not inherently have an opposite, but it can be understood as a double, which serves as a sort of opposite of the number two – i.e., four is related to number two. Only in this sense can voice be opposed, that is, related to silence.

> Therefore, since the body is a quantity and the voice is a body, it is evident that none of them is a substance. However, almost nothing is opposed to quantity, but silence is opposed to voice. This difficulty, however, we resolve through a comparison of similar cases. The number four, inasmuch as it is quantity, is opposed to nothing; but insofar as it is double, it is opposed to two through a relation. Similarly, Socrates, inasmuch as he is a human, is opposed to nothing; but insofar as he is a father, he is opposed to his son. In the same way, the voice, inasmuch as it is quantity, is by no means opposed to anything; but inasmuch as it admits of a contrary, we do not deny that it is opposed to silence. From these examples, an attentive reader could observe: everything that the soul perceives through the body, whether visible or invisible, falls under the term 'body'. [...] Hence, corporeal and visible are interchangeable, when their privations are universally considered: every visible thing is corporeal, though not every corporeal thing is visible. Hence, every incorporeal thing is invisible, though not everything invisible thing is incorporeal.[90]

It is worth specifying that Abbo's conception of relation as opposition is not entirely accurate. The nuances of this distinction become clearer when considering texts like the *Excerpta isagogarum et categoriarum*, an anonymous text, structured as a dialogue about logic, closely associated with the tradition of Gerbert of Aurillac's *De rationali et ratione uti*.[91] In the *Excerpta*, there is a refined exploration of the category of relation, emphasising that not all relative terms are necessarily contrary. For instance, some terms, like 'virtue' and 'vice', can be both related and opposite, while others, such as the 'double', lack a proper opposite but maintain a relationship with other terms.[92] This nuanced understanding provides a more comprehensive view of the dynamics within the category of relation than Abbo's simplified oppositional framework.

Beyond this inaccuracy concerning the category of relation, Abbo's arguments in support of the corporeity of the voice primarily emphasise its perceptibility and quantifiability, contributing to a long-standing debate on this topic. His position aligns with that of John Scotus Eriugena, as seen in his commentary on Priscian's *Institutions of Grammar*.[93] Eriugena, Priscian, and the Stoics share the perspective that the voice is corporeal, setting them apart from other philosophers (such as the Academics and Peripatetics), who considered the voice as an accident of the air and therefore incorporeal. Eriugena provides three reasons supporting the corporeity of the voice: it is produced by a body (the air), perceived by a bodily sense (hearing), and possesses length, width, and depth like all bodies appear to have. Abbo echoes these aspects, asserting that the voice is produced by a body, perceptible through a body, and quantifiable. Furthermore, Abbo introduces an original physical and biological theory not found in early medieval or ancient sources. According to this theory, the voice is air rendered audible by a specific liquid substance or part in the brain ("cerebri liquiditas"), purifying it from other humours. This purified and rarefied air resembles a quasi-ether, which would otherwise be perceptible through the sense of smell if not purified.[94]

The discussion on the nature of the voice needs to be contextualised within a broader discourse on corporality. Abbo posits that created reality is fundamentally determined quantitatively. Consequently, the concept of a body is inseparable from that of a tangible substance defined by quantity. As demonstrated in the case of natural compounds, numbers play a crucial role in defining phenomena such as lunar phases, human life stages, and even the distinct pitches within the musical scale. This illustrates that, according to Abbo, the voice, too, despite being invisible, is quantifiable, computable, and divisible according to number. Hence, the voice is considered a body.

Notes

1 *In Calculo*, III, 19, p. 84. It is worth noting that Calcidius defines the rules proper to a discipline as *praecepta artificialia* (*In Tim.*, 342, p. 334), hence, my choice of defining them as 'artificial'.

2 *In Calculo*, III, 16, p. 82: "Post elucubratam unitatis simplicitatem rato ordine instituit disserere de illi opposita compositione, quae est aut natura aut voluntate.

Sed quae natura sunt conposita, si augmentum sui legitima progressione capiant, nullo modo aliter solvi possunt, quam conposita fuerunt; nec augmenti motus aut naturalis progressio alia via, nisi qua crevit, ad detrimentum redit ullo modo".

3 *In Calculo*, III, 16, p. 82: "Quod in promptu est secundum phisiologos, qui humanae vitae modum ab ipso conceptionis tempore per ebdomadas dierum, mensium et annorum distribuunt, dicentes quod a die nativitatis finita quinta annorum ebdomada, id est XXXV anno, humanarum virium constituta sit meta, sicque eisdem momentis retrorsum redeat ac LXX anno, id est aliis V annorum ebdomadibus, retro sublapsa deficiat. Idem experitur in lunaris cursus incremento ac detrimento, qui secunda dierum ebdomada plenior, rursus duabus ebdomadibus deficit sui augmenti momentis die XXVIII".

4 *In Tim.*, 37, pp.85–87; *In Somn.*, I, 6, 45–81, pp. 35–46; Martianus, *De nuptiis*, VII, 738–739, pp. 11–13.

5 *In Somn.*, I, 6, 76, p. 44.

6 *In Somn.*, I, 6, 48–60, pp. 36–40.

7 *In Calculo*, III, 17, p. 82.

8 I referred to the link between the liberal arts and the columns of Salomon's temple in Chapter 3.

9 *In Tim.*, 36, p. 85; *In Somn.*, I, 6, 11, pp. 26–27; Martianus, *De nuptiis*, VII, 738, p. 11. Associating the number seven to the goddess Minerva is typical of certain literature displaying some Pythagorean tendencies; for this topic, see Robbins, "The Tradition". Calcidius delves into the value of the number seven in the context of psychogony, elucidating its seven 'limits'. These limits denote the seven numerical values that form the structural foundation of both the cosmic and human soul. Macrobius, on the other hand, focuses on the age factors of Scipio Emiliano, employing this as a pretext for an extensive exploration of the significance of numbers eight and seven. Lastly, Martianus, through the discourse articulated by the character Arithmetic before the celestial assembly, expounds upon the properties inherent in the first ten numbers. In the perspectives of these thinkers, the number seven and integer numbers more broadly were deemed carriers of arithmetical properties imbued with specific mythological or physical significance. Cf. *In Calculo*, III, 17, pp. 82–83: "Septem vero eiusdem substantiae virginitati consecratur ab hoc maxime, quia solum infra denarium nec generat nec generatur, simplici dicatur sapientiae."; cf. also Evans and Peden, "Natural Science", 118.

10 *Arithm.*, I, 19–20, pp. 40–45.

11 *In Calculo*, III, 18, p. 83.

12 *In Calculo*, III, 18, p. 83: "Quos tamen ubique obsessos circumvallant reliqui ceu adversarii, innotescentes neminem perfectum debere esse securum, qui, nisi adeptis virtutibus circumspiciat, utrimque habet quo cadat, vel cessando a bono opere vel transgrediendo terminos, quos posuerunt optimi rectores mundanae reipublicae". On the link between Abbo's vocabulary and rhythmomachy, see Peden, "Introduction", in *In Calculo*, xxx. The earliest treatises on rhythmomachy were written by Asilus of Würzburg, around 1030, and by Hermann Contract (of Reichenau), around 1040. See Evans, "The *Rithmomachia*"; Borst, *Das Mittelalterliche Zahlenkampfspiel*, 59–81; Folkerts, "*Rithmomachia*". A reproduction of the chessboard with the numerated pieces can be found in Beaujouan, "L'enseignement du *quadrivium*", 647. Moral implications could be drawn from the mathematical chessboard. For instance, just as perfect numbers can be 'besieged' in rhythmomachy, perfect men risk then to desist from right conduct or to transgress the limits imposed by good rulers. This conclusion is also consistent with Abbo's political view of Christian society of his time, which needs stability to not fall into chaos and heresy. On this, see Peden, "Unity", 163–164; for Abbo's interpretation of *Prov.* 22, 28 ("Don't cheat your neighbor by moving the ancient

boundary markers set up by previous generations"), see Mostert, *The Political Theology*, 115–116. Boethius, too, hints at the moral value of perfect numbers; cf. *Arithm.*, I, 19, p. 42: "Inter hos autem velut inter inaequales intemperantias medii temperamentum limitis sortitus est ille numerus qui perfectus dicitur, virtutis scilicet aemulator".

13 *In Calculo*, III, 19, p. 84.

14 *In Tim.*, 317, p. 313.

15 *In Tim.*, 21, p. 71–72. Calcidius also usages the four Aristotelian qualities (hot, cold, wet, and dry) to make sense of the interaction of the elements, see *In Tim.*, 13–20, pp. 65–71. See Caiazzo, "La forme"; Caiazzo, "Filosofia della natura"; and also Somfai, "The Brussels Gloss".

16 *In Somn.*, I, 6, 25–26, p. 30.

17 *In Calculo*, II, 5, p. 76: "Deus, qui numeris elementa ligat, organum cordis suis nervis attemperat". Cf. *Consolatio*, III, m. 9, 10, p. 80.

18 *In Calculo*, III, 19, p. 84: "Cuius equidem lunae speralis globositas crescit paulatim ac minuitur, [...] altera parte instar speculi claritatem admittit, quam terris refundit, altera obscurior nihil lucis recipit. Cuius rei similitudinem ad medietatem usque denigrata quaelibet soliditas speralis, utpote ovi, exprimit, si nigredinem prius obiectam oculis paulatim vertis, ut admodum nascentis lunae cornutam albedinem contempleris; demum vertendo paululum, omnes species lunae contueris usque ad plenilunium et rursus ad defectum. Nam et monoydes et dyatome et amphicirtos et panselenos utriusque hemisperii satis liquido occurrit".

19 Byrhfert, *Enchiridion*, III, 2, pp. 148–151 (especially figure no. 21). William of Conches, *Dragmaticon philosophia*, IV, 10, pp. 127–128. On Abbo's lunar diagram in William of Conches, see Obrist, "Guillaume de Conches", 185–187.

20 The diagram I am referring to is in Abbo, *Sententia*, 128 (Figure 5.3). The same diagram is reproduced in Engelen, *Zeit*, 94 on the basis of the ms. Cambridge, Trinity College, R. 15. 32 (945), fol. 5v.

21 Obrist, "Le tables", 175 See also Obrist, "Le tables", 172–177 for Abbo's lunar phases transcribed in the ms. F of Peden's critical edition, that is Berlin, Staatsbibliothek, Phill. 1833.

22 Boethius, *In Perihermeneias*, I, 2, pp. 52–59. Boethius uses the expression "secundum placitum voluntatemque".

23 This is the thesis of Suto, *Boethius*, 66–9. Different interpretation in Engels, "Origine". On name imposition, see also Rosier-Catach, "Regards croisés".

24 Boethius, *In Topica*, I, PL, 64, 1048B.

25 *In Calculo*, III, 20, pp. 84–85: "Dico autem secundum placitum esse quae voluntate fiunt vel facta sunt. Quae et ipsa non multum a natura discrepant, si rationabiliter arte constant; quoniam omnis ars imitatur naturam, profecta ex passionibus animae, quas credamus naturales esse. Hinc est quod ex tribus aequalibus terminis producuntur omnes species inaequalitatis, rursusque in eosdem resolvuntur ordine suae productionis. Nam post aequalitatem prima species inaequalitatis occurrit multiplex, secunda superparticularis, tertia superpartiens, quarta multiplex superparticularis, quinta multiplex superpatiens, per trinam praeceptorum regulam".

26 *Arithm.*, I, 22–31, pp. 46–66. Boethius' theory of inequality ratios are condensed in *In Calculo*, III, 72–80, pp. 116–122.

27 *In Calculo*, III, 73, p. 117. Cf. *Arithm.*, I, 23, 1, p. 47.

28 The ratios can be expressed in different ways. Cf., for instance, Ambrosetti, *L'eredità*, 20; and Jean-Yves Guillaumin's *nota ad locum* in his edition of *Arithm.*, I, 32, 1, p. 66.

29 *In Calculo*, III, 74, p. 117: "Est enim superparticularis, quotiens maior numerus semel totum minorem et ipsius minoris insuper unamquamlibet non plus continet partem". Cf. *Arithm.*, I, 24, 1, p. 50.

30 *In Calculo*, III, 75, p. 118: "Maior [superpartiens] sane numerus semel totum minorem recipit; sed hoc a superparticulari differt, quod non unamquamlibet, sed duas vel tres partes minoris assumit". Cf. *Arithm.*, I, 25, 2, p. 58.
31 *In Calculo*, III, 77, p. 199: "Siquidem multiplex superparticularis est quotiens secundum naturam multiplicis minor numerus maiorem bis vel ter vel quotienslibet metitur. Et ipsius minoris insuper maior unamquamlibet non plus continet partem". Cf. *Arithm.*, I, 29, 4, p. 61.
32 *In Calculo*, III, 79, pp. 120–121: "Quinta inaequalitatis species multiplex superpartiens proditur, quae ex diversitate multiplicis et superpartiens contexitur. Quae species ex duabus conpacta fit, quotiens secundum naturam multiplicis minor numerus maiorem bis ter vel quotienslibet metitur. Qui maior non unamquamlibet minoris partem sed duas vel tres vel quot ipsa tulerit conparatio prorsus assumit". Cf. *Arithm.*, I, 31, 1, p. 65.
33 *Musica*, I, 4, pp. 191–192 (the paragraph titled *De speciebus inaequalitatis*).
34 On the rhythmomachy, see above note 12.
35 *Arithm.*, I, 32, 3–7, pp. 67–68. Slightly modified translation by Masi, *Boethian Number Theory*, 114.
36 *In Calculo*, III, 22, pp. 85–86.
37 *In Calculo*, III, 20, p. 85. Cf. *Arithm.*, I, 32, 9–28, pp. 68–73. On Abbo's production rule, see also Peden, "Boethius", 169–170.
38 There are two more charts in the *Explanatio* illustrating the production, by means of the same arithmetical rule, of the double super-three-partient (*duplus supertripartiens*) and the double super-four-fifth (*duplus superquadriquintus*). Cf. *In Calculo*, III, 25–26, p. 88.
39 *In Calculo*, III, 23, pp. 86–87. Cf. *Arithm.*, II, 1, 3, p. 79.
40 *Arithm.*, II, 1, 4, p. 79. From the English translation by Masi, *Boethian Number Theory*, 123, the challenge I refer to does not emerge due to the incomplete translation of the sentence.
41 *Arithm.*, II, 1, 5–7, pp. 79–80.
42 *In Calculo*, III, 23, p. 87.
43 *In Calculo*, III, 24, p. 87.
44 Pivotal to this discussion is the study Otisk, "The Interpretations", which offer a comparison of the method of reducing inequality ratios developed by Gerbert and Notker. I refer to Otisk's study for a detailed explanation of Gerbert's and Notker's rules. See also Evans, "The Saltus Gerberti". To frame Notker's thought, I refer to Delville, Kupper and Laffineur Crepin, *Notger*.
45 Gerbert of Aurillac, *Costantino suo*, in Gerbert of Aurillac, *Epistulae*, 692–9. In all likelihood, Gerbert wrote this letter between the 972 and the 982, at the time Abbo wrote his *Explanatio*. Cf. *Gerberti Opera mathematica*, 32–35. For Gerbert's scientific works, see also Materni, "Attività scientifiche".
46 Notker of Liège, *De superparticularibus*, 297–299.
47 *In Calculo*, III, 28, p. 89: "Omne conpositum partibus est coniunctum. Partium autem coniunctio solutionem expectat aut natura aut placito. Sed natura solubile in suam originem constat redire. Quapropter omne natura conpositum in originis suae resolvitur primordium. Est etiam quiddam placito solubile, quod viget imitatione naturae. Naturalis autem solutio non abhorret a suae conexionis principio. Cuncta enim mutabilia suis momentis proficiunt et in id unde coeperant revertuntur. Eius igitur imitatione solubile contentum est suae conpositionis origine. Nam nullum conpositum caret ortu. Omne quoque solubile conpositum est. Nullum igitur solubile caret ortu. Unde constat quod nullum carens ortu solubile sit".
48 Peden, *Glosses*, vol. I, p. 105: "Quicquid enim dissolvi potest omni rerum necessitate constat esse compositum. Quod autem compositum est, simplex non est. Igitur simplex est, neque ullo modo dissolvi potest; quae ideo non recipit divisionem quia nec admittit compositionem". The manuscript can be reasonably placed

between the end of the 10th century and the beginning of the 11th. Cf. Eastwood, "Manuscripts", 146. According to Vezin, this manuscript was copied by an English scribe: Vezin, "Leofnoth", 110–111.

49 *In Somn.*, I, 5, 17, p. 23.

50 Lapidge, "Litanies", 266; Barker-Benfield, "A Ninth-Century Manuscript", 150.

51 *Gen.* 1:2 (Septuagint version: ἡ δὲ γῆ ἦν ἀόρατος καὶ ἀκατασκεύαστος). Differently, in the Vulgata, the earth is said *inanis et vacua.*

52 *In Calculo*, III, 32, pp. 91–92: "Sed legimus 'terra erat invisibilis et inconposita'. Quae procul dubio si ante discretionem sui ab aliis elementis erat quadam confusione inconposita et invisibilis; facta visibilis, conposita esse coepit, non quidem ex aliis elementis, sed ex qualitatibus suis".

53 Augustine, *De Genesi ad litteram*, I, 13, 27, pp. 19–20; Augustine, *Confessiones*, XII, 21, 30, pp. 231–232. I refer to Tornau, "Intelligible Matter".

54 *In Tim.*, 278, pp. 282–283. The four senses of matter/earth are based on the analysis by Reydams-Schils, *Calcidius*, 131–133; see also the classical study Van Winden, *Calcidius*, 170.

55 Bede, *Opera exegetica*, I, 2, pp. 5–6.

56 Remigius of Auxerre, *Expositio super Genesim*, 1, 1–2, pp. 6–9. Hrabanus Maurus, *Commentaria in Genesim*, PL 107, 445A-446A.

57 *Periphyseon*, II, 549D, 33.

58 *Periphyseon*, II, 548C, 31; cf. *Periphyseon*, III, 690D-691A, 102.

59 *Hebd.*, 187, ll. 26–27; and *Hebd.* 188, ll. 41–43. Ans also Boethius, *In Categorias*, PL 64, 200C-202A.

60 *In Calculo*, III, 32, pp. 91–92: "Nam dum omnium rerum visibilium et invisibilium quaedam causa est, cui per se id est esse quod est, ea nihil vacat in creaturis, etsi non est materialis, conponens singula suis formis, quibus valeant describi. Sicque fit ut, cum ea quae creata sunt, per aliud ac non per se subsistunt, post acceptam essendi formam simplicia non sunt; ut huic lapidi aliud est esse, aliud id quod est, quippe aliud est materialiter per rem, aliud essentialiter per substantialem qualitatem. Quocirca vera est praedicatio quae per figuram aethiologiam, id est causae redditionem, proponit, quod unumquodque idcirco divisioni necessario subiacebit, quia conpositum est, scilicet ex diversis, ut dictum est. Diversum est enim esse et id quod est. Haec quoque conpositio ornatum rebus subministrat, sine qua cuncta sunt inania, vacua, informiaque".

61 Rand, *Johannes Scottus*, 51–52.

62 *In Calculo*, III, 35, p. 94.

63 Beda, *De temporum ratione*, 5, p. 283: "Dies est aer sole illustratus, nomen inde sumens quod tenebras disiungat ac diuidat". Cf. Abbo's definition of the day, in his *Sententia*, 128, ll. 118–122: "Quamvis quidam falso arbitrentur, diem non posse dici nisi mane incipiat, et vespere desinat, idcirco quia iuxta xv speciem diffinitionis dies est sol super terram. Sed nos, ut dictum est, diei nomine xxiiii horarum spatium dicimus".

64 Cf. Boethius, *In Categorias*, PL 64, 195D-196A.

65 *In Calculo*, III, 36, p. 94. For an explanation of terms involved in the second-figure syllogism, which are intended respectively as the cause and the effect, cf. Abbo, *De hypotheticis syllogismis*, IX, p. 88: "Horum similitudine exemplorum plurima suppetit copia, quandoquidem multa sunt quae per conexionem fiunt. Aliquando enim enuntiativi termini simul sunt et alter alterius causa est, ut: 'Si sol super terram est, dies est'". Cf. also Abbo, *Sententia*, 128, ll. 121–122.

66 Cicero, *De inventione*, I, 26, 39: "Tempus autem est – id quo nunc utimur, nam ipsum quidem generaliter definire difficile est – pars quaedam aeternitatis cum alicuius annui, menstrui, diurni nocturnive spatii certa significatione". For the definition of the hour (*hora*), cf. Isidorus, *Etymologiae*, V, 29, 2, p. 89: "Hora enim finis est temporis, sicut et ora sunt finis mans, fluviorum, vestimentorum".

67 *In Somn.*, I, 21, 8–22, pp. 123–127.
68 The gloss is reported by Peden in *In Calculo*, 95, n. 111. For the ms. 7299, see the previous section; for the ms. Paris, BnF, lat. 16677, see Barker-Benfield, "A Ninth-Century Manuscript", 149–155.
69 For the water-clock in antiquity see Dohrn van Rossum, *History of the Hour*, 21–22.
70 Beda, *De temporum ratione*, III, p. 276: "Recipit autem hora iv punctos, x minuta, xv partes, xl momenta, et in quibusdam lunae computis v punctos". On Bede's technical vocabulary, see Lejbowicz, "Postérité médiévale".
71 *In Calculo*, III, 37, p. 95: "Hora, quae est finis eius temporis, quod cum sit quaedam pars aeternitatis, tamen profertur secundum certam significationem sui alicuius spatii, ut annui vel diurni. Nam cum maiores nostri, astrologiae dediti, aequinotialem diem noctemque simul clepsidra in XXIIII aequis partibus dividerent, uniuscuiusque partis finem cum aspirationem 'horam' vocaverunt, ipsisque partibus sive horis XXIIII totam machinam caeli cotidie revolvi innotuerunt. Clepsidram autem dixerunt quodlibet vas aeneum, per cuius inferiora, facta parvissimo foramine, decurrendo quasi furatur aqua, quia et 'clepere' furari, et 'ydor' aqua sunt interpretati. Cuius aquae fluxu ac diligenti receptione seu partitione, ut Macrobio auctor est, experimento probaverunt itum ac reditum totius caelestis sperae, obliquus qua se signorum ordo semper consuevit vertere, attributis duabus horis singulis XII signis. Sed et horarum spatium [...] propter subtiliorem facilitatem supputationis distribuunt in IIII vel V punctis, punctumque unum in tribus partibus".
72 McCluskey, "Gregory of Tours", 20–21. See also North, "Monasticism"; Balmer, "The Operation".
73 *Horologium stellare monasticum*, 1–18. The attribution of this *horologium* to the abbey of Fleury has been advanced by Minois, "L'Horologium stellare".
74 Thierry d'Amorbach, *Consuetudines*, 29, p. 220: "Quescentibus autem illis usque ad horam competentem custos ecclesie previsa clepsidra lento pede dormitorium ascendit et scyllam, que in medio pendere supra infantum lectulos solet, diutius movet".
75 Gerbertus, *Epistolae*, 153, pp. 375–379, for the water-clock p. 379. On this letter I will return later.
76 Cf. McCluskey, *Astronomies and Cultures*, 111–112.
77 *In Calculo*, III, 38, p. 96: "Fiebatque talis consideratio hoc modo: pretaxatum vas aeneum obturato foramine replebatur aqua. Quo diligenter praeparato, notabatur de fixis stellis quaelibet clarior una, serenitate ante praecognita. Ipsius itaque stellae ortu et apparitione reserato foramine, incipiebat mox aqua decurrere. Qui decursus aquae continuatus durabat per totam ipsam noctem et sequentem diem usque ad ipsius stellae rursus ortum et apparitionem, quae designabat totius caelestis sperae integram conversionem. Aqua vero decurrens colligebatur cum magna diligentia, ut aequaliter in XXIIII partibus distributa, in idem vas aeneum evacuatum refunderetur tota. Et post tres vel quotlibet dies sumptis duobus vasculis, quorum singula essent capacia illius aquae, tantum vicesimae quartae partis notabatur designatae stellae ortus, et statim incipiebat ipsius aquae decursus. Cuius vicesima quarta pars supposito uno ex duobus vasculis recepta, significabatur esse transacta una hora. Ne vero interim, dum de uno duorum vasculorum iactabatur aqua, saltem periret una gutta, subiciebatur aliud ut eo pleno hora secunda significaretur. Quo duarum horarum spatio unumquodlibet ex XII signis zodiaci oriebatur aut occidebat procul dubio. Sicque fiebat, ut alterna successione vasculorum agnosceretur et horarum spatium et magnitudo signorum".
78 *In Somn.*, I, 21, 15, p. 125.
79 *In Calculo*, III, 38, p. 96: "Siquidem aequinoctiali tempore medietas illius aquae tota defluebat in nocte et altera ex integro in die. Aestivo vero solstitio sex portiones illius aquae in nocte, XVIII residuae in die. Hyemali quoque solstitio

totum fit a contrario, dum in eo XVIII sunt in nocte, sex reliquae in die. Iterea quorumque dierum augmentum vel detrimentum pandebatur adiectis vel dectractis in aquae fluxu aliquibus unciis unciarumve minutiis. Cum haec ita sint, certum est diem et horam esse vocabula temporis; tempus vero, sicut et corpus, speciem quantitatis, non in eo quod est, sed in eo quod quantum est".

80 Gerbertus, *Epistulae*, 153, p. 376: "Tibi in pigno amicitiae misi quaedam ex astronomicis subtilitatibus collecta, scilicet accessus et recessus solis, non secundum eorum opinionem colligens, qui aequales fieri putant singulis mensibus, sed eorum rationem persequens, qui describunt omnino inaequales". A thorough study on Gerbert's timekeeping is Otisk, "Letter on Timekeeping".

81 Gerbertus, *Epistulae*, 153, p. 376: "Martianus quippe in astrologia incrementa horarum ita fieri putat: 'Sciendum – inquit – a bruma ita dies accrescere, ut primo mense duodecima ejusdem temporis quod additur aestate accrescat, secundo mense sexta, tercio quarta, et quarto mense alia quarta, quinto sexta, sexto duodecima'"; cf. Martianus, *De nuptiis*, VIII, 878, p. 463.

82 Cf. Boethius, *In Categorias*, PL 64, 209C.

83 *De statu animae*, I, 25, p. 88.

84 *In Calculo*, III, 31, p. 91: "Solemus equidem integritatem ei rei tribuere, quam nec fracturam nec scissuram nec diminutionem novimus habere. Quam etiam soliditatem habere diceremus, nisi hanc pro crassitudine corporis frequentius acciperemus, qua nihil caret visibile, nec etiam filum araneae. Nec mirum cum omne visibile corporeum sit, etsi non omne corporeum visibile existit".

85 *In Calculo*, III, 42, pp. 98–99: "Omne quod anima per corpus sentit, corporeum sit necesse est. Ipsa nempe utitur corporis diversis officiis velut mancipiis, per quae sensibilia, id est quae sentiri possunt extrinsecus, renunciantur sibi quasi dominae intrinsecus. Renunciantur autem interdum, ut sensibilia omnia visibilia. Omne ergo visibile corpus. [...] Quaerendum est itaque quae sint incorporea, quae numerabili divisione caeduntur. Nihil enim numero subiacet praeter corpus [...]. Porro, quod numero non subiacet, nullo modo in partes scinditur. Incorporeum quidem numero non subiacet, nec in partes ergo scinditur. Ast ipsa vox invisibilis numerorum proportionibus aut gravatur aut acuitur, ut delectabiliter audiatur. Quae si auditur, sensibilis est. Omne quoque sensibile corpus. Unde constat vocem corpus esse [...]".

86 *In Calculo*, III, 5, p. 76.

87 Cf. Boethius, *In Peri hermeneias editio secunda*, IV, 10, p. 266.

88 *In Calculo*, III, 42–43, 99: "Unde constat vocem corpus esse, quanquam contradicat aliquis huiusmodo ratiocinatione. Silentium, inquit. opponitur voci, sicut tenebrae luci. Oppositorum autem nihil substantialiter est. Corpus vero substantialiter est, ut lapis. Est enim lapis substantia. Vox igitur corpus non est. Cuius obiecta breviter enervamus, quia in talibus operam consumere nodum est in scirpo quaerere. Omnis substantia corporea ita circumdatur includiturque his, quibus substat, ut sine eis nequaquam visui subiaceat [...]. Substantia enim quantitate non eget nisi per corpus, quippe sibi sunt convertibilia quantitas et illud ut propria. Nam ubicunque quantitas est, corpus abesse non potest, et ubicunque corpus est, quantitas sit necesse est. Quare cunctis vera sapientibus notum est, corpus esse attributum accidenti, non substantiae, quoniam eius esse quantitas est proprie, solam hanc differentiam in una eademque re, verbi gratia lapide, anima intelligibiliter pervidente. Nam eidem lapidi aliud est esse, aliud quantum esse, aliud etiam quale esse".

89 Cf. Boethius, *In Categorias*, PL 64, 221D.

90 *In Calculo*, III, 42–43, 99: "Cum ergo corpus quantitas sit, vox vero corpus, constat eorum nihil esse substantialiter. Sed quantitati fere nihil opponitur, voci vero silentium oppositum est. Quam ansam calumniae dissolvimus ex similium collatione. Quaternarius, eo quod est quantitas, nulli opponitur; sed eo quod

duplus est per relationem binarius oppositus est. Socraten etiam, eo quod homo est, nulli opponimus; eo quod pater est, filio oppositus est. Eodem modo vox eo quod quantitas est, nequaquam opposita est; eo quod contrarium admittit, non denegamus eam silentio opponi. Ex quibus sobrius lector animadvertere potuit: omne quod anima per corpus sentit, seu visibile, seu invisibile sit, corporis vocabulo contineri. [...] Unde reciprocentur inter se corporeum et visibile, eorumque privationes prolate universaliter. Omne visibile corporeum; non tamen omne corporeum visibile. Hincque fit ut omne incorporeum invisibile sit, etsi non omne invisibile incorporeum existit".

91 The critical edition of this dialogue (see *Excerpta isagogarum*) has been made by Giulio d'Onofrio on the basis of three manuscripts. One of them is Città del Vaticano, BAV, Reg. lat. 1281, which, on fols. 37v-52r, also transmits Abbo's *Explanatio*.

92 *Excerpta isagogarum*, 116, p. 106.

93 On the ancient debate, cf. Aulus Gellius' mention of Democritus, Epicurus, Lucretius, Plato and the Stoics in his *Noctes Atticae*, V, 15, 6–7, p. 209. Eriugena's commentary is edited by Luhtala, "Early Medieval Commentary"; see also Mainoldi, "*Vox*"; and Rosier Catach, "*Vox*".

94 *In Calculo*, III, 5, p. 76. Mamertus (*De statu animae*, I, 7, pp. 45–46) refers to air carrying both smells and sounds to the ears and nose, respectively. However, no mention is made about the 'liquid of the brain' ("cerebri liquiditas") which can purify and elaborate the sound, as stated by Abbo.

6 Arithmetic and Calculus

6.1 Fractions: Ounces and *minutiae*

Any quantity, whether it be a solid, a liquid body, or an abstract unit of measurement, can be comprehended as a whole and subsequently divided into parts. One such example is the division into ounces (*unciae*), which represent equal portions of the 'axis', that is, the integral unit standing for whatever thing conceived as a whole.[1] Ounces constitute the pre-modern system of fractions and find historical documentation in the context of weight measurement systems.[2] Abbo would have encountered information about them in various sources, including Isidore's *Etymologies* (referred to by Abbo as *Veriloquia*), Remmius Favinus' *Carmen de ponderibus et mensuris* (a didactic poem in hexameters composed at the end of the 5th century, presenting both Greek and Latin systems of weights and measures), and an anonymous work titled *De ponderibus* (which Abbo attributes to a certain Virgil of Tolosa, potentially Virgil the Grammarian from the 7th century).[3]

Fractions of the kind recalled in the abovementioned texts, developed within a duodecimal system and were grounded on the basic unit of one-twelfth, specifically the ounce (*uncia*), which today would be represented as $\frac{1}{12}$. In essence, the system encompasses fractions with a denominator of base twelve (including multiples and sub-multiples of twelve), originating from the measure of weight where the integer number one is equivalent to the 'axis', corresponding to twelve ounces.

During the medieval period, fractions underwent a transformative process, gradually transitioning from associations with concrete, measured objects to being perceived as 'abstract parts' of a unit conceived as a whole. Ultimately, fractions came to be treated in a manner similar to numbers, devoid of a necessary reference to real-world objects.[4] This conceptual shift led to the adoption of conventional symbols to represent fractions. Figure 6.1, arranged from left to right, displays these symbols (*notae*), their Latin names, the number of parts of the axis to which they correspond, and, lastly, their corresponding modern numerator and denominator.[5]

Twelve ounces corresponds to an axis (*as*), symbolised by the Latin letter 'I', sometimes adorned with a mid-crossed comma. Each duodecimal fraction

DOI: 10.4324/9781032643472-7

I	as, assis	12	1
	deunx, iabus	11	$\frac{11}{12}$
	dextans, dextas, distas	10	$\frac{10}{12}$
	dodrans, dodras	9	$\frac{9}{12}$
	bisse, bes	8	$\frac{8}{12}$
	septunx	7	$\frac{7}{12}$
	semis	6	$\frac{6}{12}$
	quincunx	5	$\frac{5}{12}$
	triens, treas	4	$\frac{4}{12}$
	quadrans, quadras	3	$\frac{3}{12}$
	sextans, sextas	2	$\frac{2}{12}$
	sescuncia	–	$\frac{1}{8}$
	uncia	1	$\frac{1}{12}$

Figure 6.1 The axis and its ounces. Table based on *In Calculo*, 140.

derives its nomenclature based on its relationship either with the axis or the ounce: names ending in *-as* indicate the number of parts relative to the axis, while those concluding with *-unx* denote the number of ounces.[6] In the *Explanatio*, Abbo employs the term 'proportion' (*proportionalitas*) to characterise the relationship between fractions and the axis or the ounce. This concept of proportion is drawn from Boethius' *De arithmetica*, where it signifies the connection between integer numbers. Boethius had distinguished ratio (*proportio*) from proportion (*proportionalitas*): the former denotes the correlation of two numbers (e.g., 2:4, a double ratio), while the latter signifies the correlation of two ratios, encompassing at least three distinct numbers (e.g., 1:2 = 2:4, expressing a double proportion).[7] Abbo extends the technical notion of proportion to fractions. According to him, it is conceivable to identify a proportion involving four numbers between the ounces and the axis. Each specific fraction can be viewed not only as a part or portion of the axis itself but also as constituted by several equal aliquot parts or ounces. This proportion, encompassing both ounces and the axis, is precisely articulated by the names assigned to the fractions. For instance, the *treas* consists of four ounces, signifying four equal aliquots of the axis ($\frac{4}{12}$) and, consequently, representing the third part of the axis ($\frac{1}{3}$).

> It should be noted that the endings of words in *-as* or *-unx* indicate specific quantities, as those ending in *-as* represent an axis, and those in *-unx* designate an ounce. For example, *treas* indicates a third of an axis, and *quincunx* denotes five ounces. [...] However, it must first be explained what the difference is between ratio and proportion. Ratio is the comparison of any two numbers to each other, such as 2 to 4; while proportionality involves three or more numbers, like 2, 3, 4; and a ratio never consists of more than two terms, nor proportionality of less than three terms. [...]

> Single parts [of the axis] compared to the total sum [*i.e.*, the axis], which, as mentioned, is divided into twelve equal parts, are denominated according to their comparison. Since 12 is six times in proportion to 2, four times in proportion to 3, and three times in proportion to 4, names are formed from this proportionality, such as two ounces [*i.e.*, $\frac{2}{12}$] are called sextans, three ounces *quadrans*, four ounces *triens* or *treas*.[8]

Ounces are not the smallest conceivable subdivisions of the axis. Abbo introduces the concept of *minutiae*, which are additional aliquots of the ounces and are designated by distinct symbols (see Figure 6.2). The *calcus* stands as the smallest aliquot of the ounce, originally denoting a minute stone which, within the measurement system, equated to the weight of two grains of wheat. Two *calci* constituted one *cerates*; four *calci* comprised one *obolus*; and eight *calci* formed one *scripulus*. According to Isidore, a *scripulus* is equivalent to the weight of either 16 lentils or 18 grains of barley. The *scripulus* is further divided into six *siliquae*, constituting half of a *dimidia sextula* and a quarter of a *sextula*. Within an ounce, there are 24 *scripuli*, while in an axis, there are 288. Abbo provides comprehensive explanations for these conversions in the passage of the *Explanatio* reported below.

> The term 'minutia' refers to *calci*, *cerates*, *oboli*, and all smaller weights from which half of the ounce is composed. The smallest and first of these weights is called *calcus*, which, as mentioned, is a tiny stone to be balanced with two grains of lentils, and when doubled, it forms *cerates*. A doubled *cerates* makes an *obolus* [...]. Two *oboli* also make a *scripulum*, consisting in the weight of six *siliquae*. Therefore, according to the

Γ	uncia	$\frac{1}{12}$
ʃ	semuncia	$\frac{1}{24}$
UU	duae sextulae *o* sesclae	$\frac{1}{36}$
Ɔ	sicilicus	$\frac{1}{48}$
U	sextula, sescla	$\frac{1}{72}$
✱	drachma, dragma	$\frac{1}{96}$
Ψ	dimidia sextula *o* sescla	$\frac{1}{144}$
ℌ	scripulus	$\frac{1}{288}$
÷	obolus	$\frac{1}{576}$
Z	cerates	$\frac{1}{1152}$
℘	siliqua	$\frac{1}{1728}$
o~	calcus	$\frac{1}{2304}$

Figure 6.2 The *minutiae*, i.e., aliquot parts of the ounce. Table based on *In Calculo*, 140.

> *Veriloquia* by Isidore, a *scripulum* is balanced with 16 grains of lentils, although Virgil of Tolosa in his writings claims that 18 grains of barley are weighed, counting three grains for each *siliqua*. Finally, a doubled *scripulum* eventually makes a *dimidia sextula*, and when this is doubled, it equals a *sextula*, that is, a whole *sescla* [...]. As the ounce increases gradually, the number of *scripuli* grows, so that an axis has twelve ounces, and each ounce has 24 *scripuli*, making a total of 288 within the entire axis.[9]

In the *Explanatio*, technical notions such as those mentioned above are introduced to elucidate the multiplication of fractions of the ounce and simplify arithmetical operations involving quantities representing the smallest parts of one unit. As illustrated in Figure 6.2, these operations may encompass fractions ranging from $\frac{1}{24}$ up to $\frac{1}{2304}$. Abbo devises a method to streamline these calculations, relying on the conversion of the multiplicand fraction of ounce into the integer count of the corresponding *scripuli*. Subsequently, the number of *scripuli* can be converted into ounces and smaller fractions. To illustrate this, consider the following example: to multiply a *sicilicus* by ten, the initial step involves converting the *sicilicus* into the corresponding *scripuli* (1 *sicilicus* = 10 *scripuli*). Next, it is necessary to multiply the integer count of the *scripuli* by ten (6 × 10 = 60). Finally, the product of the *scripuli* is converted into the respective ounces and smaller fractions (60 *scripuli* = 1 *sextans* + 1 *semuncia*). Abbo adeptly transitions between the duodecimal and decimal systems during these conversions. Figure 6.3 provides an exemplification of the conversion between ounces and *scripuli*.

In the context of multiplying ounces, Abbo presents an alternative and more intuitive procedure alongside the method of conversion into *scripuli*. This method begins with the explicit assumption that each ounce represents

as	I	288 ℈
deunx	[symbol]	264 ℈
dextans	[symbol]	240 ℈
dodrans	[symbol]	216 ℈
bisse	[symbol]	192 ℈
septunx	[symbol]	168 ℈
semis	[symbol]	144 ℈
quincunx	[symbol]	120 ℈
triens	[symbol]	96 ℈
quadrans	[symbol]	72 ℈

Figure 6.3 Equivalence of ounces and *scripuli*. Table based on *Calculus*, 52–53.

an aliquot of a specific whole quantity divisible into 12 equal parts. From this standpoint, if an ounce is multiplied n-times, it surpasses the given whole quantity by x. The remainder x can then be translated into the corresponding quantity of ounces. To illustrate this approach, consider the multiplication of the *sextans* by ten. The *sextans* ($\frac{2}{12}$) is equivalent to two ounces (i.e., two equal aliquots of the given axis). When multiplied by ten, it results in twenty ounces. Given that the axis corresponds to twelve ounces, the surplus from the multiplication equates to eight ounces. Consequently, the twenty ounces can be readily understood as one axis (1) and one *bisse* (8 ounces), representing 1 and $\frac{2}{3}$.[10] To a contemporary reader, these examples may indeed appear strikingly straightforward and fundamental. However, they offer insights into a crucial phase in the historical development of mathematics, revealing a still rudimentary yet significant method for performing calculus and arithmetic in the Latin world.

6.2 Representing Integer Numbers

Abbo's exploration of duodecimal fractions, along with his examination of integer numbers and their various representations, constitutes an important chapter in the evolution of mathematical thought during this period. As anticipated in the very first chapter of this book, in the context of the historical evolution of mathematics, Abbo's understanding of arithmetic can be temporally situated between the advent of the abacus in the Latin world and the establishment of the Indo-Arabic numerical system.[11] During Abbo's time, diverse systems for representing numbers were available: one based on written symbols and another employing specific parts of the human body, known as the Roman and 'finger' numeral systems, respectively. The Roman numeral system, characterised as 'additive', entails expressing numbers through the arrangement of specific letters in a series, with each letter possessing a distinct value that is either added or subtracted from others in the series to derive the represented number. In contrast, the 'finger' system involves representing numbers by configuring specific figures using the fingers themselves. Both systems adhered to a decimal base and lacked the concept of zero. Evidently, neither can be classified as a positional system strictly speaking. While the Roman system allows for the subtraction or addition of a digit's value based on its relative position to other digits, each digit (or letter) maintains a fixed value. Unlike a positional system, the value of each digit in the Roman system does not result from multiplication by a position-specific factor. Nevertheless, as I explore in the subsequent sections of this chapter, Abbo's illustrative multiplication methods suggest an incipient acknowledgment of the positional system, providing evidence of its developmental trajectory.

In addition to the Roman and 'finger' methods of representing numbers and performing calculations, we have already seen that Abbo utilised another system of symbols standing for fractions organised on a base 12. Being widely employed since antiquity, particularly for measuring weight, this system existed alongside the Roman and 'finger' ones. Abbo's instructional approach in arithmetic and calculus encompassed the integral ten-based numbers of the Roman, 'finger' systems, and the symbols for duodecimal fractions. To these, another one can be added which links arithmetic and calculus to geometry.

To elucidate the multiplication of integers, Abbo relies on another representation system grounded on Boethius' *De arithmetica*. There, Boethius outlined various methods of representing numbers. These included the Roman notation, where a letter signifies a number (e.g., V for five); the linear notation, where units are sequentially added on the same line (e.g., 8 represented as III-IIIII); and the 'figured numbers', where numbers are symbolised by plane and solid geometric shapes like triangles, squares, pentagons, hexagons, pyramids, parallelepipeds, etc. This technique originally had connections to the representation by arrays of points, tiny stones, or pebbles. Equilateral figures are formed by evenly distributing points along their sides, while quadrilateral figures with unequal sides result from placing varying quantities of points along the sides. More specifically, numbers expressible through tetragons encompass all squared natural numbers: $1^2 = 1$; $2^2 = 4$; $3^2 = 9$; $4^2 = 16$; and so on. These numbers generate plane geometric figures with equal sides, with the length corresponding to the radix of the represented numbers.[12]

Abbo's discussion also aims to highlight a specific parallelism between the multiplication of integers and that of geometric figures. Notably, he demonstrates that a particular arithmetical regularity observed in the multiplication of even and odd numbers is mirrored in the construction of specific geometric shapes. Regularity and consistency Abbo refers to are of this kind: when an even number is multiplied by another even number, the result is a greater even number; this holds true for odd numbers as well, as the multiplication of two odd numbers always produces a greater odd number. Or also, the multiplication of an odd number by an even number yields a greater even number. This regularity is also evident among plane figures. A quadrangle, representing a square plane figure formed by a square number, when multiplied by a similar one, generates another tetragon with equal sides. Similarly, a rectangle, characterised by one side longer than the other by one unit (*altera longior*), will produce a similar figure when multiplied by another rectangle. When a quadrangle is multiplied by a rectangle, it results in another rectangle, albeit one with a side longer than the other by more than one unit (*ante longior*).

> The first substantial distinctions of all number must be examined, namely, even and odd. If one of them, namely, even, grows by the multiplication of any even number, it continually remains even. However, if an odd number multiplies an odd number, the product does not deviate from the nature of the odd. If an even number also increases by the quantity of an odd or an odd by that of an even, [the product] never becomes odd, but always remains even. These distinctions between numbers are consistent with certain plane figures, namely, squares and those with one side longer than the other. For if a square multiplies a square, a square is produced. Similarly, if a rectangle is multiplied by another, a rectangle arises. Yet, if

> one is extended by the multiplication of the other [*scil.* if a square is multiplied by a rectangle], neither a square nor a rectangle (*altera longior*) is produced; instead, a rectangle with a side longer than the other of at least more than one unit always results (*ante longior*).[13]

With this foundation regarding the products of odd and even numbers, it becomes clear that a distinctive feature of Abbo's arithmetic and calculus is to elucidate the relationships between numbers across different representational systems. In the subsequent sections, I will observe Abbo's techniques to show how the calculus (and more specifically, multiplication) can be performed: by fingers and by tables.

6.3 Calculus by Fingers or the *loquela digitorum*

The technique of calculating with fingers has its origins in ancient Egypt, and evidence of its use can be traced in both Greek and Latin literature.[14] During the Middle Ages, this method was extensively elucidated in the first chapter of Bede's *De temporum ratione*. In the chapter titled *De computo vel loquela digitorum*, Bede provided a detailed explanation of how to count units and tens, as well as hundreds and thousands, using fingers (*digiti*) and their joints (*articuli*). A straightforward illustration of this method is given in Figure 6.4, based on manuscript Vatican City, BAV, Urb. lat. 290, fol. 31r, precisely transmitting Bede's *De temporum ratione*.

The *digiti* in this context refer to the middle, ring, and little fingers, while the *articuli* correspond to the thumb and index finger. In this finger-counting system, units (1–9) and hundreds (100–900) were counted on the *digiti* of the left and right hand, respectively. Meanwhile, tens (10–90) and thousands were counted on the *articuli* of the left and right hand, respectively. While this system is clearly not positional, Faith Wallis notes that there is an implicit positional representation of numbers, inasmuch as the fingers were assigned values based on the powers of ten.[15] Using this method with two hands, one could count and perform arithmetical operations up to 9999 on a decimal base. The calculation involved forming various figures by combining both the *digiti* and the *articuli*. Abbo adopts and describes this technique as follows:

> The explanation of all these things will become clearer if the progression of multiplied numbers is traced back from the smallest numbers [*scil.* by units], recognizing that the rationale of multiplication is suitable for the fingers. Indeed, units are assigned to the bended fingers of the left hand, whereas tens are assigned to joints of the fingers of the same hand. Fingers bended touching their roots [*i.e.*, their metacarpal] are: the little finger for one, the ring finger for two, and the middle finger for three. However, the same fingers can bend downwards towards

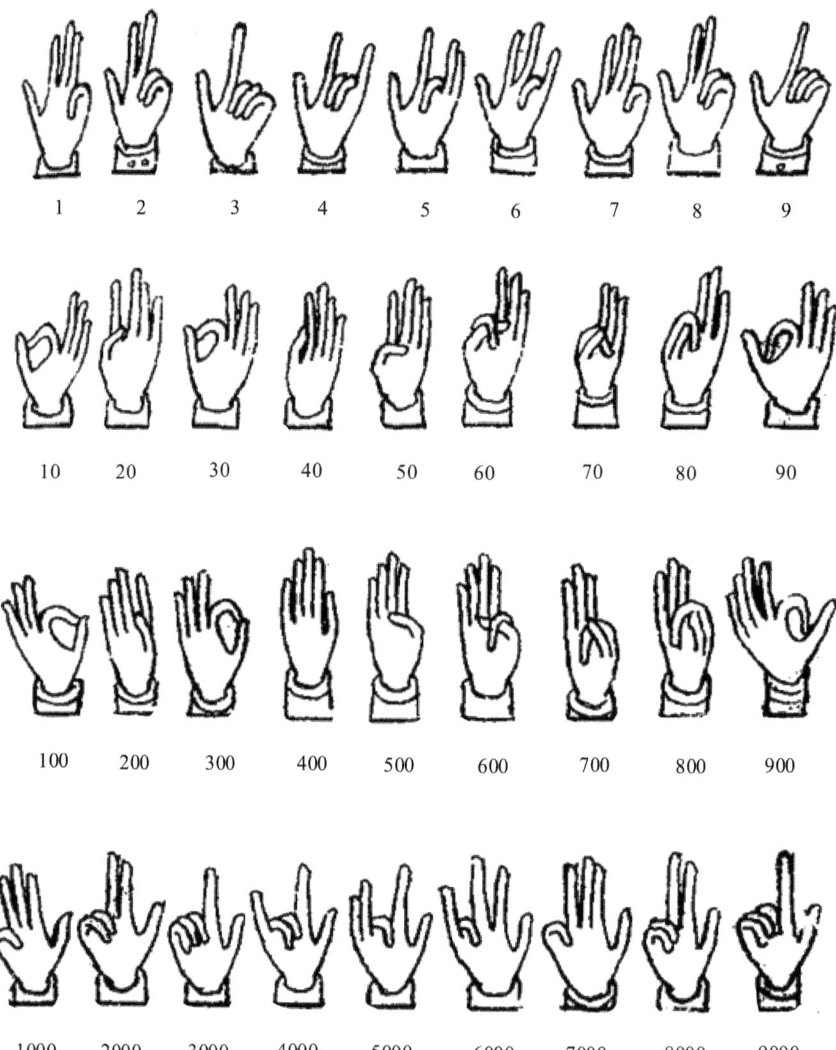

Figure 6.4 Fingers positions for numbers 1-9000. Cf. Bede, *The Reckoning of Time*, 256 and Vatican City, BAV, Urb. lat. 290, fol. 31r.

the base of the palm: the little finger [stands] for seven, the ring finger for eight, and the middle finger for nine. Of these, to represent the number four, the little finger extends while the other two bend; to represent the number five, only the middle finger is bent while the other two are extended. To represent the number six, the middle and the little fingers extend, while only the middle bends towards the metacarpal. The thumbs and index fingers represent tens, either clasping each other at their joints or overlapping one another.[16]

Abbo's procedure for representing the product of multiplication with fingers focuses on only one hand, specifically, the left hand. The rationale for presenting an example involving only the left hand stems from Abbo's method, which aims to simplify multiplication between integers. He suggests reducing the multiplicands to their respective units or *singulares* (for example, from 20×20 to 2×2), resulting in a product consisting of no more than tens and units. These can be precisely represented with the left hand alone. Once the (quite easy) mental calculation is done (for instance, 2×2=4), the result is then represented using the fingers by lifting up the little and index fingers as well as the thumb of the left hand. Subsequently, the same fingers and joints are assigned precise values according to the powers of ten inherent in the multiplicands. Abbo illustrates this rule with an example, demonstrating the multiplication of 60 by itself.

> If you multiply ten by ten, you will assign [the value of] 100 to each finger and of 1000 to each joint. For example, when you calculate 60 multiplied by 60, you find that six times six is 36, where three are [to be assigned to] joints and six to fingers, indicating that 60 multiplied by 60 is 3600. For 30, the thumb and index finger touch each other in a clasp, which stand for three thousand, consistently with the previous explanation. This method of multiplication must always be observed, first carefully noting what numerical values fingers and joints are assigned. For if you multiply units by tens, you will assign the value of 10 to each [finger] and the value of a 100 to each joint.[17]

As Abbo explains, if one were to count 60 times 60, the procedure would start by dividing 60 by 10 and reducing it to its respective unit, which is six. Subsequently, six is multiplied by itself, yielding 36. The representation of 36 with the fingers involves the thumb and index finger joined like forming a circle, symbolising three tens, while the little and middle fingers are lifted up, and the ring finger is bent towards the palm, representing the six units. Given that the multiplication of 60 by itself involves the tens, the joints are assigned the value of one thousand, and the digits are assigned the value of hundreds. By adding these two values, the final product of the multiplication is obtained, which is 3600.

6.4 Calculus by Tables

In addition to the calculus by fingers, Abbo presents another method for multiplying integer numbers, namely, by using tables. The first table presented by Abbo (Figure 6.5) in the *Explanatio* is inspired by a table by Boethius and also corresponds to the one found in the *Enchiridion* by Byrthferth, who was Abbo's most well-known disciple. Another version of the same table appears in the Oxford manuscript, St. John's College, 17 – I shall call it the 'Ramsey table'.

L A T I T U D O

L	1	10	100	1 000	10 000	100 000
O	2	20	200	2 000	20 000	200 000
N	3	30	300	3 000	30 000	300 000
G	4	40	400	4 000	40 000	400 000
I	5	50	500	5 000	50 000	500 000
T	6	60	600	6 000	60 000	600 000
U	7	70	700	7 000	70 000	700 000
D	8	80	800	8 000	80 000	800 000
O	9	90	900	9 000	90 000	900 000

Figure 6.5 Table for multiplication of the five powers of ten by the first nine natural numbers based on *In Calculo*, 110.

The table Abbo presents consists of six columns, referred to as *latitudo*, and nine rows, labelled as *longitudo*. As has just been said, Abbo closely follows a table presented by Boethius in his work *De arithmetica*. Although the structure of both tables is fundamentally the same, there are slight differences in content due to variations in Boethius' and Abbo's intentions. Boethius arranges the first ten natural numbers on both the upper side of the board (*latitudo*) and the left side (*longitudo*). The subsequent rows display the products of the numbers placed in the first row and the first column. This arrangement results in the square numbers (i.e., 4, 9, 16, 25, and so on up to 100) being positioned along the diagonal of the table.[18] Boethius aims to demonstrate that the ratio of multiples precedes all other ratios. In other words, he seeks to show that the relationship between two multiple numbers is simpler than any other type of arithmetical inequality ratio. On the other hand, in Abbo's table (Figure 6.5), the numbers are distinct from Boethius' arrangement. In the first row, Abbo places the number one followed by the first five powers (*unitatis naturae*) of ten (i.e., 10^1, 10^2, 10^3, 10^4, 10^5). The first column contains the first nine natural numbers (1–9).[19] The subsequent rows display the products of the powers of ten multiplied by the nine natural numbers. Consistently with Abbo's henology, the table demonstrates that the first power of ten (10^0, namely, 1) is the source of all the numbers and remains identical to itself when multiplied by itself. In contrast, the other powers of ten, when multiplied by themselves, transform into the next larger power. For example, ten times ten (10^2) produces one hundred, the next power in the series, and so forth for the others. The connection with Boethius' table and objective becomes evident when Abbo converts the same table into a diagram illustrating the inequality ratios that link the products of the powers of ten and the first nine natural numbers. This diagram will be discussed below. For now, it suffices to note that Abbo takes Boethius' table, which illustrates the emergence of inequality ratios, as a model, while altering its contents to eventually demonstrate (in another diagram) that natural numbers raised to the powers of ten preserve the same kind of inequality ratios.

As mentioned earlier, Abbo's table also corresponds to the one reported in Byrhtferth's *Enchiridion* – a text serving as an introduction to Byrhtferth's computus, written in both Latin and Anglo-Saxon. Byrhtferth places Abbo's table in the section of his work (Book IV) that deals with number symbolism (that is, arithmology) and the 'ages' of the world. Although the numbers reported in both tables are identical, Byrhtferth's interpretation of the table differs significantly from Abbo's. Byrhtferth focuses on the table as indicative of the perfection of the number 1000, yet without specifying the reason for such perfection.[20] The number itself and the table bear a pure symbolic value in Byrhtferth's interpretation. In contrast, when presenting the table in the *Explanatio*, Abbo refrains from introducing arithmological remarks and sticks to the arithmetical notions implied. Although Byrhtferth explicitly refers to Abbo as an "expert in mathematics and perfect in philosophy" right below his table, it appears that Byrhtferth reinterprets Abbo's arithmetical teaching concerning this table independently and in accordance with his own interest in arithmology.[21]

Another table that establishes a connection between the *Explanatio* and Abbo's arithmetical teaching at Ramsey is found in the manuscript St John's College, 17, on folio 35r. This manuscript, dating back to the 12th century, originated from Thorney Abbey, located not far from Ramsey – this is the reason why I refer to the table in this manuscript as 'Ramsey table'.[22] It is a miscellany on time reckoning with complementary materials concerning medicine, mathematics, and cosmology. It also preserves Abbo's acrostic poem in honour of St. Dunstan and *Ephemerida*, as well as excerpts of his astronomical treatises. As in Byrthferth's case, the numbers in Ramsey table are identical to those in Abbo's table. However, what sets the Ramsey table apart is that it is accompanied by two couplets and is followed by multiplication rules attributed to Abbo. These rules are the reason why the Ramsey table is often referred to as 'Abbo's abacus' by modern scholars.[23] Neither the Ramsey table nor Abbo's table in the *Explanatio*, though, are technically an abacus: for an abacus is meant to perform arithmetical operations utilising the positional system, whereas the tables I am examining merely display the powers of tens.

The couplets at the top of the Ramsey table state: "In hac figura descriptus est numerus infinitus: incipit enim ab uno pervenitque usque ad nongentesimum millesimum", which translates to "In this figure, an infinite number is inscribed: for it goes from 1 up to 900,000". While it is evident that both Abbo's and the Ramsey table do not represent an infinite number, there must have been a reason for the specific limitation to numbers up to 900,000. According to Charles Burnett, Abbo's decision to limit the numbers to 900,000 is attributed to the symbolic value of the number of columns drawn in the table, which is six.[24] As mentioned in the previous chapter and in line with Boethian number theory, the number six is considered a perfect number in arithmetic.[25] However, I have already established that Abbo's table does not imply any arithmological concerns. A more plausible explanation for this limit can be found in the practice of calculating by fingers, which, in Abbo's *Explanatio*, runs parallel to the practice of calculating by tables. To understand both why the Ramsey table is supposed to represent the infinite number and why it actually stops at 900,000, one need to refer to the works of Boethius, Bede, and Martianus Capella. The notion that numbers proceed infinitely finds its roots in Boethius' *De arithmetica*: unlike magnitude, which can be indefinitely divided, multitude can increase infinitely.[26] In addition to this assumption, the possibility of producing numbers greater than 900,000 is supported by the method of calculating with fingers, as explained by Bede in his *De temporum ratione*. Bede refers to the *Romana computatio*, a treatise on the ancient Roman technique of finger reckoning and a catalogue of numerical signs from the late 7th century. Bede argues that it is possible to count up to 1,000,000 precisely by using not only the fingers but also other parts of the body, such as the neck, chest, belly, hips, and so forth.[27] However, in limiting the multiplied numbers to 900,000, Abbo (and the copist of the Ramsey table) might have been influenced by the advice given by the character of Arithmetic in Martianus Capella's *De nuptiis Philologiae et Mercurii*. In the Book VII of this work, Arithmetic recommends using only the fingers for calculation purposes and avoiding excessive contortions of the body, as such movements are deemed unsuitable for philosophers. Philosophers, according to Arithmetic, must adhere to a specific order in their bodily movements and reject anything indeterminate (what is infinite), which, in a broader sense, is not suitable for scientific pursuits.[28] This statement by Arithmetic is explicitly mentioned in the *Explanatio*.

> Indeed, the [character of] Arithmetic in Martianus' [work] declares that "only that number is deemed proper which can be constrained by the fingers, as otherwise", she says, "a certain distorted dance of the arms occurs". For example, when we represent 90,000 by seizing the left thigh with the left hand and turning the thumb toward the groin, and for a 1,000,000, clasping both hands together, we somehow imitate the gestures of dancers – thus, the infinite multitude of numbers is considered, which is rejected by philosophers, who delight in comprehending, through science, the multitude of finite quantity.[29]

Beside this statement concerning the philosophers' rejection of indefinite quantities, the reason for not inserting another column in the table is that the first five powers of ten are sufficient to achieve the table's purpose: showing how numbers are multiplied by themselves and indicating the consistent arithmetical links and ratios among them. Moreover, as Abbo explains, the table suffices as it is to show that the links between the numbers hold also on a grammatical level, if one considers the names assigned to the multiplied numbers. In each row of the table, except for the first, we find numbers such as the tens (*decenes*), the hundreds (*centenes*), the thousands (*millenes*), etc., whose names are derived from those of the units (*singulares*) arranged along the first column. A clear example of this 'grammatical relation' is observed in the case of the number twenty (*viginti*), resulting from multiplying ten by two. As elucidated by Abbo, the etymology of the name 'twenty' (*viginti*) reveals its connection with its respective unit, the number two: *viginti* stems from *bis genitus*, meaning 'generated twice' or *bis-ginti*, from which its name, *viginti*, is derived.

> Certainly, each row of the latitude-side is marked with numbers that derive their names from the specific units, except for the first [line], which is distinguished by different principles and is subject to the nature of unity. For example, from *binarius* [i.e., two], [derive the names] *biginti* (which is said 'born twice of ten'), two hundred, two thousand, twenty thousand, two-hundred thousand. In the second row of the latitude-side, the origin of the name [is shown], so that it is clear to what extent it is useful also for other [rows]. The number two multiplied by itself yields four. Twenty multiplied by itself results in the multitude of four hundred. Two hundred multiplied by itself grows to forty thousand. In the same way, three times three is nine, and thirty times thirty is nine hundred, and three hundred times three hundred is ninety thousand. Do you not see here how the single rows follow one another, so that their totals are evident in the table, when from four arise forty and forty thousand, of which one comes from twenty, the other from two hundred?[30]

In addition to the grammatical consistency underlying the relationships between numbers, there is another arithmetical link between factors and products. This link becomes evident when identifying the same arithmetical inequality ratios that were illustrated by Boethius' table in the *De arithmetica*. As mentioned earlier, Abbo develops his own table based on Boethius', altering its contents to highlight the special status of unity and its connections with other powers of ten. Despite the numerical differences between the two tables, Abbo maintains Boethius' objective by translating the table into another diagram, demonstrating the inequality ratios among the products of the powers of ten and the first nine natural numbers. Abbo's diagram (Figure 6.6) consists of five columns connected by arcs. The first column includes the powers of 10, starting with 1 (10^0). The second column represents the number two and its respective products when multiplied by the powers of ten. The same order is followed for the number three in the third column, and so forth for the numbers four and five. Abbo's diagram appears as if he had reoriented the previous table, placing the numbers along the *longitudo* side at the top from left to right and those along the *latitudo* side from top to bottom. This diagram shows that the ratios identified among units are consistent with the ratios among the respective tens, hundreds, and thousands.

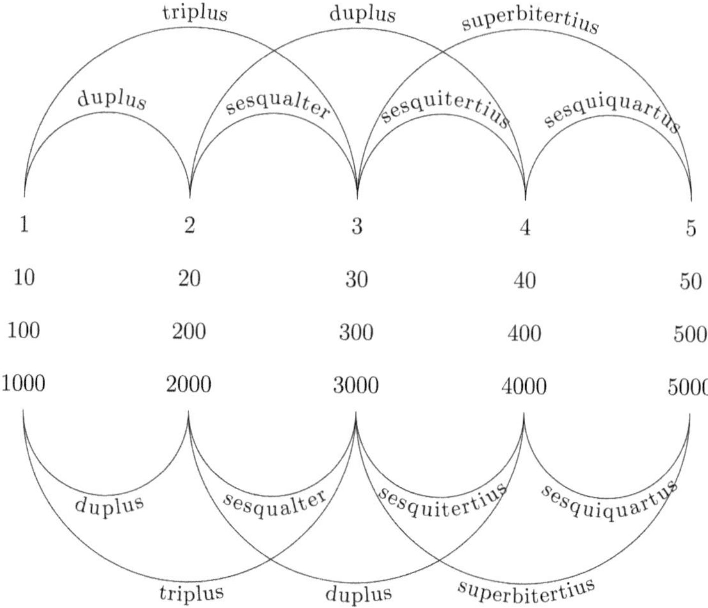

Figure 6.6 Diagram of inequality ratios based on Bamberg, Staatsbibliothek, Class. 53, fol. 34r.

In this diagram, the inequality ratios among units, as well as tens, hundreds, and thousands, are clearly exemplified. Specifically, it demonstrates that just as two is double one, or four is double two, similarly, 20 is double 20, and 40 is double 20. This pattern continues with 200 being double 100, and 400 being double 200, and so forth in the case of the thousands. Likewise, three is sesqualter of two, and 30, 300, and 3000 are sesquialter of 20, 200, and 2000, respectively. All the ratios between the first five natural numbers – double (2:1, 4:2), triple (3:1), sesqualter (3:2), sesqui-third (4:3), sesqui-fourth (5:4), and super-two-thirds (5:3) – are preserved in the corresponding tens, hundreds, and thousands.

6.5 Multiplication Rules

Abbo provides specific rules for the multiplication of numbers. These rules align with the principles mentioned earlier regarding finger reckoning and partly correspond to those found in the St. John's College manuscript below the Ramsey table. While expressing the same underlying principle, the rules in the *Explanatio* apply to calculations supported by table (Figure 6.5), whereas those in the Oxonian manuscript refer to finger reckoning. However, both articulate a straightforward method for calculating the products of units, tens, hundreds, and thousands multiplied by each other ("ad invicem collatos"): the multiplicands (tens, hundreds, or thousands) are first reduced to their respective units, then the product of the units is calculated, and finally, the result is assigned a precise value according to the powers of ten. For instance, when multiplying 20 × 20, the tens are first reduced to units, the result (2 × 2 = 4) is calculated, and, since it involves multiplication between tens, the value of 100 is assigned to the product (4 × 100 = 400). In calculus by finger, the value of hundreds would be directly assigned to the fingers representing 4, while in calculus by table, the process involves identifying the row corresponding to the unit (4) and sliding along it until crossing the column of the specific power of ten (hundreds). These multiplication rules are detailed in the *Explanatio* as follows.

> Certainly, if you are unsure how much twenty times twenty or twenty times thirty amounts to, consider that twice two is four, and twice three is six. And since you are investigating the multiplication of tens by tens, assign the value of a hundred to each unit of the multiplied sum – you will find that twenty times twenty is 400, and twenty times thirty is undoubtedly 600. [...] If however, you multiply units by tens or tens by units, the same method with smaller numbers should be applied in accordance with the aforementioned rule, counting ten for each unit. For instance, twice twenty and twice thirty first are meant as twice two and twice three, for the latter couple produces four and six, the former forty and sixty. [...] If you multiply tens by hundreds or hundreds by tens, you must follow the aforementioned method with smaller numbers, taking a thousand for each unit. For example, if you inquire about the result of twenty times 200 or twenty times 300, it will become evident that twice two and twice three result in four and six, so that the former results in 4,000 and the latter in 6,000. Similarly, if you multiply units by hundreds and hundreds by units, through the previous method with smaller numbers, you will attribute a hundred to each unit. For that five times 500 is 2,500 is clear from the fact that five times five is twenty-five. Again, if you multiply hundreds by hundreds, you will calculate ten thousand for each unit. For instance, three times three, which produces nine, reveal that 300 times 300 is 90,000. [...] If you multiply units by thousands or thousands by units, remember to attribute the value of one thousand to each unit of the smaller product. For everyone knows that six times a 1,000 or a 1,000 times six is 6,000. [...] If you multiply thousands by tens or tens by thousands, you will make sure that you have substituted 10,000 to each unit of the smaller product; like in the case of four, which is twice two, shows that twenty times 2,000 is 40,000. If you let the thousands grow by multiplying them by hundreds or the hundreds by thousands, you will give a hundred thousand to each unit of the smaller product. For we know that 200 times 3,000 is 600,000, because we know that twice three is

> six and we do not neglect to attribute the value of a hundred thousand to the six units. If you attempt to produce another sum by multiplying thousands by thousands, following the aforementioned rule for the smaller numbers, ensure that you compute a million [for each unit]. For if you inquire about how many are 2,000 times 2,000, you will draw from the example of twice two, which is four. Attributing the value of a million to the units of the number four, you will know that 2,000 times 2,000 is 4,000,000. However, since this pertains to the method of the abacus, it belongs to a different discussion and task.[31]

Abbo's rhetorical prose does not present the multiplication rules in a schematic and easy-to-memorise manner, unlike the Oxford manuscript where they are laid out as a litany, serving as a mnemonic device.[32] While Abbo's rules and those below the Ramsey table represent two different methods of multiplying numbers (using the table and the fingers, respectively), they align in adopting the general principle of assigning a specific corresponding value according to the powers of ten. This ambivalence is also evident in other texts focusing on arithmetical operations. For instance, these multiplication rules are found in a letter written by Gerbert of Aurillac to Constantine of Fleury and in the *Ratio numerorum abaci* by Heriger of Lobbes (d. 1008).[33] Both texts embody the same arithmetical principle applied to calculus by fingers and by table, respectively. As in the case of the Oxford manuscript, Gerbert emphasises the precise value assigned to the fingers of the hand in order to calculate the product of tens multiplied by hundreds: "If you multiply tens by hundreds, assign to each finger (*digitus*) the value of ten, and to each joint (*articulus*) the value of a hundred".[34] On the other hand, Heriger seems to refer to the position of a finger in relation to an unspecified space, likely a column on a multiplication table: "If a unit is multiplied by another number, place the finger (*digitus*) on the same [spot]".[35] This approach is also evident in another anonymous text on the abacus, known as the Echternach abacus, preserved in manuscript Trier, Stadtbibliothek, 1093/1694.[36] It includes a series of texts on calculation, including multiplication rules, designed to determine the column in which the product of the multiplication falls. This abacus, along with the associated notes on calculation, traces back to the 10th-century abbey of Echternach and is influenced by Gerbert's teaching. In the *Explanatio*, Abbo presents multiplication rules specifying that he will not go into details that precisely pertain to calculus by abacus. The table analysed (Figure 6.5) is not an abacus, even though it shares the division into columns. It substantially differs from the abacus due to the distribution of the powers of ten. In the abacus, these powers are arranged in ascending order from right to left.[37] Conversely, in Abbo's table, there is an inverse arrangement of the same powers of ten, increasing from units, from left to right, up to hundreds of thousands. This specific arrangement also allows Abbo to emphasise once more the role of unity as the origin of all numbers, even in such a technical context.

Notes

1. *In Calculo*, III, 47, p. 101.
2. See Maher and Makowsky, "Literary Evidence".
3. The title *Veriloquia* used by Abbo to refer to the *Etymologies* is also documented in his logical works; see Abbo, *De syllogismis hypoteticis*, IX, p. 100. Isidore, *Etymologiae*, XVI, 25, pp. 258–276; Remmius Favinus, *Carmen de ponderibus*, 29–37; and the anonymous *De ponderibus*, 138–140.
4. This passage in the history of mathematics is retraced in Menninger, *Number Words*, 445–468.
5. Figure 6.1 summarises what Abbo states in *In Calculo*, III, 50–51, pp. 104–105. Fraction symbols are called *notae* by Abbo, who also specifies that they are not by nature but rather by convention; see *In Calculo*, III, 50, p. 105.
6. Cf. Chapter 2, specifically the *Preface* of Victorius' *Calculus*, where the *particulae* or ounces are introduced.
7. *Arithm.*, II, 40, 2–4, p. 140. The inequality ratios presented in the previous chapter and expressed by a set of three numbers are in fact proportions consisting of two ratios. On the meaning of *ratio* in the medieval mathematical thought, oftentimes mistaken for proportion, see Crialesi, "Les « raisons »".
8. *In Calculo*, III, 47–48, p. 102: "Advertendique sunt fines vocabulorum in –as vel in –unx desinentium, quoniam quae in –as assem demonstrant, quae in –unx unciam designant. Treas nempe tertiam assis, et quinqunx quinque uncias innotescit. [...] Prius tamen quid inter se differant proportio ac proportionalitas dicendum est. Proportio est duorum quorumlibet numerorum ad se invicem collatio, ut II ad IIII; proportionalitas vero trium vel plurium, ut II, III, IIII; nunquamque proportio constat plus terminis duobus, nec proportionalitas minus tribus, integrum. [...] Partes singulae ad totam summam, quae ut dictum est in XII aequas portiones distribuitur, conparatae denominantur. Siquidem XII proportionaliter ad II sextuplus, ad III quadruplus, ad IIII triplus est. Ex qua proportionalitate finguntur nomina, ut duae unciae sextans, tres quadrans, quattuor triens vel treas dicantur".
9. *In Calculo*, III, 52, pp. 105–106: "Minutias dicit calcos, cerates, obolos et omnia minora pondera, ex quibus constituitur unciae medietas, quorum primum et omnium minimum dicimus calcum, qui est, ut diximus, lapis parvissimus appendendus lentis granis duobus, qui duplicatus facit ceratem. Duplicatus cerates constituit obolum [...]. Duo quoque oboli faciunt scripulum, qui perficitur pondere sex siliquarum. Quapropter secundum librum Veriloquiorum Isidori scripulus appenditur XVI granis lentis, licet Virgilius Tholosanus in suis opusculis asserat pensari XVIII grani <h>ordei, annumerans tria grana singulis siliquis. Denique duplicatus scripulus facit tandem dimidium sextulae, quod dimidium duplicatum reddit sextulae, id est sesclae [...] Post vero crescentibus paulatim unciis, usque ad assem crescit numerus scripulorum ita, ut assis habeat XII uncias et unaquaeque uncia XXIIII scripulos, quo fiunt in summam assis CCLXXXVIII". Cf. Isidore, *Etymologiae*, XVI, 25, 8–12, pp. 262–266; and *De ponderibus*, 138, ll. 24–27.
10. *In Calculo*, III, 56, p. 108.
11. For this particular moment in the history of mathematics and the development of numerical systems, see Folkerts, "Names and Forms".
12. *Arithm.*, II, 4–27, pp. 88–120.
13. *In Calculo*, III, 58, p. 109: "Primum totius numeri substantiales differentiae inspiciendae sunt, videlicet par et impar. Quarum altera, id est par, si paris cuiuslibet multiplicatione concrescat, par continuo exuberat. Si vero impar imparem multiplicet, multiplicatus ab imparis natura non discrepat. Si par quoque imparis vel impar paris quantitate extuberescat, nunquam impar, sed semper par provenit. Quas numeri differentia sequuntur quaedam planae figurae, tetragoni scilicet ac

parte altera longiores; quia si tetragonus tetragonum multiplicet, tetragonus producitur, et si parte altera longior parte altera longiorem, parte altera longior exoritur. Si vero alter alterius multiplicatione extendatur, nec tetragonus, nec parte altera longior sed ante longior semper producitur".

14 On the Egyptian origin of this technique, see Ifrah, *Histoire universelle*, vol.1, 132–138. Textual references to the same technique in the Latin world are gathered in Williams and Williams, "Finger Numbers". Cf. also Chapter 1.

15 Bede, *The Reckoning of Time*, 255. In the same volume (256), Faith Wallis provides a useful illustration of the figures formed by the fingers based on the manuscript Vatican City, BAV, Urb. lat. 290, fol. 31r. The same illustration is also reported in Figure 6.4.

16 *In Calculo*, III, 67, pp. 114–115: "Quorum omnium planior erit expositio, si multiplicatorum repetatur a minimis progressio, perspecto quoniam digitis praefixa multiplicationis ratio convenit. Siquidem singulares reflexis sinistrae manus digitis, deceni attribuuntur ipsius articulis. Reflectuntur autem ad suas radices intrinsecus: minimus pro unitate, medicus pro binario, impudicus pro ternario. Ad radicem vero palmae deorsum brachium versus recurvantur idem ipsi digiti: minimus pro septenario, medicus pro octonario, impudicus pro novenario. Quorum minimus pro quaternario, incurvis duobus extenditur; pro quinario, solo impudico iacente, medicus et minimus eriguntur. Pro senario quoque, impudico et minimo erectis, solus medicus medio palmae defigitur. Decenorum etiam pollex et index sunt indices, se invicem articulatim aut amplexantes aut superinplentes".

17 *In Calculo*, III, 68, p. 115: "Si multiplicaveris decenum per decenum, dabis unicuique digito C et omni articulo mille. Verbi gratia, dum sexagies sexagenos perquiris, sexies senos XXXVI esse invenis, ubi sunt tres articuli et sex digiti, qui ostendunt sexagies sexagenos esse III milia DC. Nam pro triginta pollex et index deosculantur se amplexu blanso, quod pro tribus milibus fit praetaxato indicio. Quem modum multiplicandi ubique observare necesse est, diligenter praecognito quot inprimuntur digitis, quot singulis articulis. Nam si multiplicaveris singularem numerum per decenum, dabis unicuique decem et omni articulo C".

18 *Arithm.*, I, 26, 2, p. 54.

19 *In Calculo*, III, 59, p. 110.

20 Byrthferth, *Enchiridion*, IV, 1, p. 228, figure no. 35. In all likelihood, Byrthferth bases his arithmological considerations about this number on what Gregory the Great and Bede stated about it. Cf. Gregorius Magnus, *Moralia in Iob*, 35, 16, 42; Beda Venerabilis, *De templo Salomonis*, II, 787.

21 On Byrhtferth's characterisation of Abbo as expert in mathematics, cf. what has been said in Chapter 1.

22 The link to the folio of the digitised manuscript where the Ramsey table is to be found is the following: https://digital.library.mcgill.ca/ms-17/folio.php?p=35r&showitem=35r_5ComputusTablesTextsII_32MathSAbboAbacus. The Ramsey table is also edited in *Gerberti Opera mathematica*, 203. A thorough description of the Oxford manuscript can be found in Wallis, *Oxford St. John's College 17*, 393–394.

23 The details of these multiplication rules will be discussed in the next section.

24 Burnett, "Abbon de Fleury", 131–132.

25 Cf. Chapter 5, the section dedicated to the arithmological value of natural compounds.

26 *Arithm.*, I, 1, 6, p. 8.

27 Beda, *De temporum ratione*, 1, 270–271. Cf. *Romana computatio*, 106–108; and Cordoliani, "A propos".

28 Martianus, *De nuptiis*, VII, 746, p. 16: "Mihi vero solus numerus approbatur qui digitis coercetur; alias quaedam bracchiorum contorta saltatio sit". On the inappropriateness of the infinite to the human understanding, see also *Arithm.*, I, 1, 6, p. 8: "Nihil enim quod infinitum est vel scientia potest colligi vel mente comprehendi".

29 *In Calculo*, III, 69, p. 115: "Siquidem Arithmetica Martiani profitetur, quod sibi 'solus numerus approbatur, qui digitis cohercetur; alias' inquit 'quaedam brachiorum distorta saltatio fit', quippe, dum propter XC milia sinistrum femur sinistra manu ita conprehendimus, ut pollicem ad inguina vertamus, atque pro decies centenis milibus ambas sibi invicem manus conplicamus, saltatricum gesticulationem aliquo modo imitamur, ubi infinita numerorum congeries perpenditur, quae a philosophis repudiatur, qui finitae quantitatis multitudinem scientia conprehendere gaudent".

30 *In Calculo*, III, 61, p. 111: "Singulae nimirum latitudinis lineae insignitae sunt numeris, qui sortiuntur vocabula ex singularibus antepositis praeter primam diversis principiis distinctam et tamen unitatis naturae obnoxiam: binarius nanque viginti, qui dicti sunt quasi biginti, is est denario bis geniti, et duecentorum et duum milium et viginti milium et ducentorum milium. In secundo versu latitudinis origo est vocabuli quod in aliis quam sit utile ita poteris advertere. Binarius in se ipso multiplicatus quaternarium reddit. Viginti quoque in se letantur quadringentorum pluralitate. Ducenti etiam in seipsis ducti excrescunt in milibus quadragenis. Eodem modo ter terni sunt novem et tricies triceni DCCCC et trecenties trecentenis XC milia. Videsne singulas versuum lineas quomodo se sequantus hinc inde perspectis summulis in figura, dum a quaternario deriventur quadringenti et XL milia, quorum alter ex XX, alter processit ex CC?".

31 *In Calculo*, III, 63–64, pp. 112–114: "Si certe ignoras vicies viceni aut vicies XXX quot sint, vide quia bis binis IIII et bis terni VI sunt et quoniam decenum per decenum investigas, contra unamquamque unitatem multiplicatae minoris summulae centum constitue, ac si invenies vicies vicenos esse CCCC, viciesque XXX procul dubio DC. [...] Si vero singularem per decenum aut decenum per singularem multiplicaveris, ex minoribus exemplum non deerit iuxta ordinem praefatae rationis, supputatis X pro unitatibus singulis. Nam bis vicenos et bis trigenos bis bini et bis terni prorsus innotescunt, dum isti in IIII vel VI, illi in XL seu LX exundant. [...] Si etiam decenum per centenum aut centenum per decenum multiplices, praefatam de minoribus regulam observare necesse est, mille acceptis pro unitatibus singulis; ut si quaeris quot sint vicies CC, aut vicies CCC, manifestabunt bis bini et bis terni per IIII et VI, quod alter IIII milia, alter VI milia restituat. Et si singularem per centenum aut centenum per singularem multiplices iuxta praefatam de minoribus tramitem, C pro unitatibus singulis accipies. Nam quod quinquies D sint II milia D, certum est ex quinquies quinis, qui sunt XXV. Rursus, si centenum per centenum multiplices, X milia supputabis pro unitatibus singulis. Siquidem trecenties CCC esse XC milia ter terni innotuerunt, qui in VIIII excrescunt. [...] Si multiplicaveris singularem per millenum aut millenum per singularem, singulis unitatibus minoris numeri M dare memineris. Nam sexies mille sive milies senos VI milia esse nemo nescit. [...] Si millenum per decenum aut decenum per millenum multiplicando auxeris, singulis minoris numeri X milia te substituisse pro certo tenebis, ut vicies II milia demonstrat esse XL milia quaternarius, qui est binarius genitus. [...] Si millenum per centenum aut centenum per millenum excrescere feceris, pro minoris numeri unitatibus singulis C milia aggregabis. Ducenties nanque III milia idcirco DC milia esse scimus, quia bis ternos VI esse conperimus ac singulis senarii unitatibus C milia tribuere non neglеximus. Si millenum per millenum in alteram summam producere conaris, pro minoribus praedictam regulam observans, decies C milia conputare curabis. Nam si quaeris quot sint bis milies bina milia, ex bis binis trahes auctoritatis copiam, qui redundant in IIII. Cuius quaternarii singulis unitatibus decies C milia tribuens, agnoscesbis milies II milia esse quadragies C milia. Sed quoniam haec pertinet ad rationem abaci, alterius sunt disputationis ac negotii".

32 To contextualise such mnemonic exercise, I refer to Carruthers, *The Book*.

33 Gerbertus, *Epistulae*, *Costantino suo Gerbertus scolasticus*, 1, 662–677; Heriger of Lobbes, *Ratio numerorum abaci*. The work by Heriger that I refer to by the title *Ratio numerorum abaci* consists in fact of succinct and different treatises, among which the one that Bubnov calls *Aliae regulae Herigeri in abacum* (*Gerberti opera mathematica*, 205–209). On Heriger, see also Verbist, *Duelling with the Past*, 15–34.

34 Gerbertus, *Epistulae*, 666: "Si multiplicaveris singularem numerum per decenum *dabis* unicuique digito decem et omni articulo centum". Italics is my choice to stress the verb by which Gerbert indicates the operation of assigning the numerical value to a finger. The verb is different from that used by Heriger, who differently states to *locate* the finger on an unspecified place (supposedly a column of a table).

35 Herigerius, *Ratio numerorum abaci*, 224: "Singularis quemcumque multiplicet, in eodem digitos *ponet*". Italics is mine.

36 Burnett, "The Abacus".

37 For Gerbert's abacus, see Folkerts, "Frühe Darstellungen".

7 Physics Before the *Physics*

7.1 The Early Medieval Concern for Natural Phaenomena

It has been a widely shared belief in 19th-century scholarship that a truly scientific interest in natural phenomena during the Middle Ages did not emerge until the early 12th century. More specifically, some scholars posited that prior to this period, the conception of nature was primarily symbolic and religious.[1] It is unquestionable that some early medieval scholars conceived of nature as a direct manifestation of God's will, thus aligning mostly with an Augustinian perspective. From this standpoint, nature was akin to a 'book written by God' – alongside the Bible, the actual book of God. Consequently, hermeneutical tools employed in the interpretation of the Bible, notably the allegorical method, influenced the understanding of nature as well.

Such a perspective on early medieval science appears overly narrow, as it fails to consider the diverse ways and trajectories through which scientific thought evolved throughout the Middle Ages, especially preceding the 12th century. Indeed, before the introduction of Aristotle's works on natural philosophy into the Latin West through Arabic translations, certain early medieval authors engaged in scientific investigations of the physical realm, distancing themselves and their research from merely symbolic interpretations of nature. This disposition, characterised by Charles Burnett as 'physics before (Aristotle's) *Physics*', has been arguably demonstrated by scholars who focused on early medieval astronomy and the physics of the elements.[2]

As highlighted by Nadja Germann, the pre-Aristotelian phase of Latin thought reveals two distinct stages in the evolution of the physics pertaining to the heavens, that is, astronomy as an autonomous discipline. The initial stage is characterised by a 'quantitative' approach to the analysis of nature, particularly evident in the 11th-century literature dedicated to astronomy and time reckoning. This quantitative approach involves, for instance, the investigation of natural celestial phenomena through a geometric paradigm. In contrast, the subsequent stage is distinguished by a 'qualitative' approach, delving into a deeper understanding of sublunar physics by considering the fundamental elements and properties that underlie natural phenomena. Germann identifies Abbo as a prominent thinker aligned with the first approach,

DOI: 10.4324/9781032643472-8

substantiating this assertion by referencing his methodology for calculating the planetary positions in relation to the zodiac, as articulated in the *Sententia de ratione spere*. Conversely, Adelard of Bath and William of Conches are recognised as exemplifying the second approach, characterised by a qualitative exploration of the foundational elements and properties governing natural phenomena.[3]

Irene Caiazzo has advocated for the existence of specific approaches to the examination of the physical realm, and more precisely, the sublunary realm, during the early Middle Ages, and delineated three primary methodologies based on late antique and early medieval literature.[4] The first approach, rooted in Neopythagorean principles, conceptualises the elements as solid numbers (i.e., cubes), drawing from Boethius' number theory. The second approach intricately weaves together mathematics and physics, assigning distinct physical qualities to each element, with these properties interconnected through mathematical proportions, as elucidated by Calcidius in his *Commentary on the Timaeus*. The third approach is fundamentally grounded in the analysis of the connections (oftentimes called *syzugiae*) among the four Aristotelian qualities (hot, cold, wet, and dry), as articulated by Macrobius in his *Commentary on the Dream of Scipio*. Of these methodologies, the most representative authors and texts are: the *Excerptum de quattuor elementis* (an anonymous diagram accompanied by a short text, which were transmitted alongside Cassiodorus' *Institutiones* and were possibly grounded on an anonymous Neoplatonic commentary on Plato's *Timaeus*), Remigius of Auxerre's and a certain magister Menegaldus' commentaries on Boethius' *Consolation of Philosophy*, Adalbold of Utrecht's commentary on the *O qui perpetua* (i.e., on the carmen from *Consolation of Philosophy* III, m. 9), and John Scotus Eriugena's *Adnotationes in Marcianum*.

In addition to these perspectives, Caiazzo highlights another facet of the early medieval physics conveyed through works that address nature-related inquiries by emphasising the role of physiological properties and incorporating observations that extend, for instance, to the biology of the human body.[5] This philosophical and literary tradition is represented, for instance, by Latin translations of Priscian of Lydia's *Solutiones ad Chosroem*. This type of elementary physics has Peripatetic roots, as exemplified by the passages of the *Solutiones* that rely on Aristotle's *De generatione et corruptione*.[6]

In this chapter, I contend that Abbo also aligns with this sort of peripatetical approach, specifically rooted in the tradition of the pseudo-Aristotelian *Problemata Physica*. Abbo's observations regarding specific physical phenomena in his *Explanatio* provide evidence that, by the end of the 10th century, the physical realm was considered worthy of philosophical speculation, albeit in its nascent 'scientific' form. This perspective gains support when examining the sources upon which Abbo bases his reasoning concerning the physical realm. On one hand, he relies on ancient and early medieval texts, such as Pliny the Elder's *Historia naturalis* and Isidore of Seville's *Etymologiae*. On the other hand, he engages with scientific literature, tracing back to the pseudo-Aristotelian *Problemata Physica*. Although records of this text were scarce during Abbo's time, he could have accessed it through Book VII of Macrobius' *Saturnalia*. As I explore further, Macrobius' *Saturnalia* and, consequently, the pseudo-Aristotelian *Problemata* serve as the foundation for Abbo's explanations regarding the different weight of oil and honey. Abbo delves into their natural and specific qualities and properties, extending his analysis to include considerations related to human physiology. In light of this, one could include also Abbo among those who adopted a 'qualitative' approach, as Germann calls it; so that one could somewhat reconsider the primacy of a 12th-century 'Renaissance' of a more scientific view of nature. It is crucial to recognise that in the *Explanatio*, Abbo does not systematically organise his reflections on physics, nor does he stick to a specific disciplinary method, such as the one developed in the 12th century *secundum physicam*. However, one could at least acknowledge that the transition to the 'qualitative' analysis of the sublunar world typical of 12th-century thinkers occurred in fact gradually, and the scientific understanding achieved by William of Conches, for instance, must be viewed as the culmination of a process that commenced decades earlier – possibly with characters like Abbo.

7.2 Scientific Literature in the Early Medieval Fleury Area

To trace the scientific background of Abbo's reflections on nature, our examination focuses first on the dissemination of specific manuscripts transmitting texts related to medicine and physics in the vicinity of Fleury. While not all the theories expounded in these texts are explicitly referenced in Abbo's *Explanatio*, I will briefly highlight some of these works to offer an insight into the state of scientific knowledge in the Northern region of today's France around the turn of the first millennium.

Concerning medical texts, two notable codices, Paris, BnF, lat. 7028, and Bamberg, Staatsbibliothek, Cod. Med. 1 (commonly known as the Lorsch Medical Book), hold particular relevance.[7] Both manuscripts were incorporated into Otto III's library, having been presented to him by the monk John Philagathos, an advisor to Empress Theophanus who later assumed the role of anti-pope under the name John XVI. The Parisian manuscript, compiled in the late 10th century, is a miscellany that includes a Latin translation of Galen's *De dinamidiis*, Celsus' *De medicina*, an excerpt from Muscio's *Gynaecia*, the Hippocratic *Liber cyrurgie*, and Vindicianus' *Epitome altera*. On the other hand, the Bamberg manuscript, of French origin and dating from the first half of the 9th century, serves as an anthology of medical Latin texts. Notably, it includes Vindicianus' *Epistula ad Pentadium* and Antimo's *De observatione ciborum* (a dietary calendar from the 6th century AD). Most crucial for our analysis, it contains the pseudo-Aristotelian *Problemata physica*, to which we will return later. These manuscripts from Otto III's library provide evidence of a continuity of medical knowledge from Late Antiquity to the early Middle Ages. Furthermore, they underscore the ongoing interest in medical theories within the Ottonian milieu, to which Abbo himself contributed and took part.

A more substantial basis for inferring the dissemination of medical texts in the geographical area of Fleury is found in manuscript Paris, BnF, lat. 9332.[8] This miscellaneous codex of medicine dates from the 8th and 9th centuries and, as noted by Augusto Beccaria, comprises a selection of the three most extensive medical treatises by Greek authors then available in Latin. Specifically, the codex puts together the Latin versions of the *Synopsis ad Eustathium* and the *Euporista ad Eunapium* by Oribasius, a brief excerpt from Celsus' *De medicina*, the *Practica* by Alexander Trallianus, and Dioscorides' *De materia medica*. It is likely that, at the beginning of the 11th century, this manuscript came into the possession of the chapter of Chartres cathedral. Some of the monograms in the margins are attributed to Fulbert of Chartres, whose interest in medicine becomes evident in certain of his epistles.[9]

In addition to medical treatises, another form of scientific literature circulated in the early medieval West, particularly in the French region. This scientific literary tradition can be traced back to the Greek genre of ἐρωταποκρίσεις, or texts structured in a question-answer format. One noteworthy example is Pseudo-Soranus' *Quaestiones medicinales*. Translated from Greek during Late Antiquity, this text comprises questions and answers on medicine, incorporating methodological and epistemological remarks.[10] Evidence of the circulation of this text in Northern France is found in two anthologies of medical texts, preserved in the manuscripts Chartres, Bibliothèque municipale, 62, and Paris, BnF, lat. 11219, dating from the 10th and 9th centuries, respectively.[11]

Additionally, the *Quaestiones salernitanae* represent another instance of this scientific literary genre. This work, associated with the Salernitan medical school, widely circulated in the Latin West especially between the 11th and 12th centuries. The text evolved from an original core, later augmented with new material from various sources: it was partially derived from Aristotelian and pseudo-Aristotelian works, covering a range of topics in physics, physiology, medicine, zoology.[12] The influence of the Salernitan medical knowledge in Northern France is commonly identified in Adelard of Bath's *Quaestiones naturales* and William of Conches' *Dragmaticon*. However, as noted by Paul Kristeller and Brian Lawn, the connection between the Salernitan medical centre and Northern France dates back to the end of the 10th century, as attested by Richer in his *Historiarum libri*.[13] There, Richer recounts a dispute on medicine (botany, pharmacology, and surgery) between a Frenchman named Deroldus and an anonymous Salernitan. This dispute, occurring before the French king (presumably Charles III), aimed to determine which of them had a superior knowledge of the subject. Deroldus is portrayed as skilled in theory, while the Salernitan possesses more practical experience.[14] Beyond the reliability of Richer's account, this story attests to Salerno's reputation as a notable centre of medicine, where a practical-oriented conception of medicine was cultivated. It also highlights the contacts between the South-Italian centre and Northern France at that time, indicating that Salerno's renown as a hub of successful medical practitioners reached France by the end of the 10th century. It is noteworthy that Richer himself mentions leaving Reims and relocating to Chartres around 991 to pursue the study of medicine, primarily based on Hippocrates' *Aphorisms* and a book titled *De concordia Yppocratis, Galieni et Surani.*[15]

Belonging to the same literary question-answer genre are the *Solutiones ad Chosroem* by Priscian of Lydia. In the early 6th century, at the multilingual and cosmopolitan Sassanid court of Persia, the Neoplatonic philosopher Priscian composed a concise question-answer compilation aimed at addressing the inquiries of the monarch Chosroes I regarding nature, psychology, and medicine. Originally written in Greek, the text has reached us through a Latin translation, most likely the work of John Scotus Eriugena, as demonstrated by Marie-Thérèse d'Alverny and Anca Dan.[16]

The last example worth considering is the pseudo-Aristotelian *Problemata Physica*. The original Greek work comprises a collection of 262 questions and answers, drawing on theories from Aristotelian and pseudo-Aristotelian texts, as well as Theophrastus' and Alexander of Aphrodisias' works. Additionally, part of its content is based on certain medical theories from Hippocrates and Galen.[17] The initial Latin translation, known as the *versio vetustissima*, covers only a limited number of issues (specifically 43 questions with their corresponding answers). This translation, which is preserved in the aforementioned Bamberg manuscript (the anthology of medical texts) among other codices, spread from Southern Italy to Northern France, more specifically Chartres.[18]

As previously mentioned, echoes of the pseudo-Aristotelian *Problemata Physica* can be discerned in Abbo's *Explanatio*, made possible through Macrobius' *Saturnalia*, which can be viewed as an instance of question-answer scientific literature to some extent. In this extensive work, Macrobius stages a lengthy dialogue among educated citizens of the Roman aristocracy.[19] *Saturnalia* is an encyclopaedic work with a dedicated scientific section, specifically Book VII. Through the character Disarius, a Greek doctor practicing in Rome, Macrobius expounds numerous physical theories based on ideas from Aulus Gellius' *Noctes Atticae*, Plutarch's *Quaestiones conviviales*, and the pseudo-Aristotle's *Problemata physica*.

Access to Macrobius' *Saturnalia* seems to be rare during Abbo's time. Our understanding of the transmission and reception of Macrobius' work is based on two primary components: indirect reports by medieval thinkers and the manuscript tradition.[20] Regarding the former, spanning from the 530s (the time of *Saturnalia*'s composition) to the Carolingian era, only two references to *Saturnalia* exist, attributed to Cassiodorus and an anonymous Irish author. Cassiodorus makes reference to *Saturnalia* V, 21, 18, employing it to comment on *Psalm* 10 and delve into the etymology of the word 'chalice' (*calix*).[21] In a distinct approach, the anonymous Irish author reworked a condensed form of *Saturnalia* I, 12–15, which centres on the Roman calendar.[22] Notably, records of Macrobius' *Saturnalia* in early medieval works are confined to the 6th and 7th centuries, but these references do not relate to physical or medieval subjects. From this perspective, to the best of my knowledge, Abbo emerges as a singular figure who both studied and utilised Book VII of *Saturnalia* before the 12th century, specifically preceding Adelard of Bath's *Quaestiones naturales*.

The *Saturnalia* are absent from the catalogue of Fleury's library dating 1552.[23] However, the manuscript tradition indicates their circulation in France, with the oldest extant exemplars dating back to the 9th century originating from this region. For instance, manuscript Paris, BnF, lat. 6370, which preserves glosses by Lupus of Ferrières and Heiric of Auxerre to Macrobius' *Commentary on the Dream of Scipio*, attests to the presence of Macrobius' work in the circle of Lupus, as the opening section of the *Saturnalia* has been copied onto the final folio.[24] The task becomes more intricate when attempting to trace the specific tradition of Book VII. Only a few existing copies preserve it alongside all the other books. One such manuscript is Paris, BnF, lat. 6371, originating from the Champagne region and dating to the early 11th century.

Despite the extensive use of physical notions and concepts from Book VII of the *Saturnalia* in the *Explanatio*, Abbo does not explicitly mention his source. This omission suggests that Abbo may have been unaware of the authorship of this text. The text of Book VII is, in fact, transmitted anonymously and in an anepigraphic way in manuscript BnF, lat. 7412, from the 12th century – raising the possibility that Abbo had access to a similar manuscript. The question of how Abbo gained access to these sources, particularly the *Saturnalia* and the *Problemata physica*, remains unanswered. Nonetheless, the acknowledgment of the spread of this scientific literature serves as a background for the analysis of Abbo's observations about natural phenomena, as further elucidated in the subsequent sections.

7.3 Qualitative Physics: From Astronomy to Female Physiology

The 'physics' section of the *Explanatio* is dedicated to addressing a particular query: the rationale behind the differing weights of two substances of the same kind. Abbo's response aligns with a form of physics that I termed 'qualitative', and it also incorporates considerations that remount to the Peripatetic tradition of the question-answer works. Throughout this section, Abbo maintains a consistent approach to the topic: he introduces a general principle and supports it with key examples that demonstrate the principle's validity within the natural realm. The underlying principle posited by Abbo is that qualities play a consistent and decisive role in natural phenomena, spanning from astronomy to physiology.

Looking at Abbo's exploration more closely, the explanation for the varying weights of two substances of the same kind lies in the interaction of the four physical qualities. More specifically, the interplay of the elements of which substances are composed depends on fundamental qualities, namely, hot, cold, wet, and dry. These qualities combine into pairs of opposites (hot-cold and wet-dry), ultimately determining the harmonious arrangement of the four elements.

> Why are things of the same kind heavier than others? And the reason is indeed evident, supported by established principles, since the diversity of the four elements is attuned through the four well-known qualities of which they consist. Cold, hot, wet, and dry, are those [qualities] which, when combined with each other, can never adhere to a contrary – by nature, those things related to lightness repel with all force the heavier ones.[25]

The interaction of qualities serves as the underlying cause for the harmony among the elements, reflecting on the overall arrangement of the cosmos. Furthermore, Abbo posits that the lightness and heaviness of bodies are contingent upon the qualities of heat and cold, respectively. Specifically, a higher degree of heat corresponds to greater lightness, while increased coldness results in greater heaviness. According to Abbo, this principle governs the structure of the sublunar world. In this respect, Abbo aligns with the Plinian perspective, wherein the tension between light (*levia*) and heavy (*gravia*) bodies plays a pivotal role in determining the arrangement and cohesion of elements in the cosmos. In this model, earth and water, being the heaviest bodies, form the dense core of the universe, holding the lighter elements (i.e., air and fire) in place and preventing their dispersion. Conversely, the lighter bodies do not collapse within the universe, as the heavier ones maintain equilibrium by exerting an upward force. The mutual and balanced interaction of opposing elements allows each to maintain its position within the cosmos.[26] While Abbo adheres to the Plinian cosmological model, he emphasises the significance of the qualities inherent in the four elements. Remarkably, Abbo highlights that the heat of air and fire tempers the coldness of earth and water. This dual influence ensures that, on one hand, the earth remains compacted and balanced on all sides by the surrounding sky, and on the other hand, the density of water is modulated by the presence of heat.

> For each thing is lighter to the extent that it participates in the essence of heat, and it becomes heavier to the extent that it is colder. Indeed, who is unaware that the heat of fire and air counteracts the coldness of water or earthly mass? That is to ask: who is unaware that the earth is balanced evenly on all sides by the surrounding sky, which encompasses it entirely each day, while water, being denser, is supported by the tempering state of the sky?[27]

Abbo's objective extends beyond the mere observation that warmer substances tend to rise while colder ones descend. Instead, he endeavours to demonstrate that, within a spherical cosmos, the qualities interact in a way that maintains equilibrium. This concept implies that at the centre of the sublunar world lies the coldest and most compact element, which is the earth. The very density of the terrestrial environment, according to Abbo, is a consequence of its cold nature. Abbo supports this correlation with an example also mentioned by Aristotle in his *Meteorology*: beneath the northern circle, water thickens due to the cold, transforming into a solid state akin to stone, namely, ice. Conversely, when exposed to a heat source, such as the sun, water reverts to its original liquid state.[28]

The impact of the cold quality also extends into the realms of mineralogy, physiology, and astronomy, as illustrated by the cases of lead and the planet Saturn. Drawing primarily from Pliny, Abbo observes that lead, being the heaviest of metals, counteracts heat: for this reason, athletes seeking to restrain their libidinal impulses use lead by placing it on the back, in correspondence of their kidneys.[29] Furthermore, Abbo contends that Saturn is the coldest among the planets, aligning this qualitative aspect with the planet's weight. The mass of Saturn, he argues, contributes to its slow orbit, taking two and a half years to complete only the twelfth part of the ecliptic.[30]

> And indeed, whatever is hardened by frost becomes denser, and it grows denser the farther it is removed from heat. Hence, under the Arctic Circle, icy water turns into stone, which is called 'ice', and it is never drawn upward by the heat of the sun to be poured back unless it has softened under appropriate warmth. Therefore, what wonder is there if certain earth are condensed by cold and, when condensed, becomes heavier than others, just as some water is pressed downward by perpetual frost? For it is well known that lead, which is naturally colder and heavier than other metals, is used by athletes, who place it on their kidneys to restrain lust. Moreover, Saturn, the farthest of the planets, due not only to the length of its orbit but also to the magnitude of its coldness, moves so slowly, completing only one-twelfth of the zodiac in nearly two and a half years.[31]

Numerous natural phenomena, spanning both terrestrial and astronomical domains, find their roots in the interplay between the qualities of cold and heat. In concluding the analysis, Abbo shifts focus to the interaction of the other two opposing qualities: wet and dry. Similar to the principles governing cold and heat in determining weight, the wetter a substance is, the heavier it becomes. Conversely, the drier it is, the lighter. In contrast to earlier explanations that drew from naturalistic theses presented by Pliny, Abbo's arguments supporting this principle are directly derived from observable facts or, notably, female physiology. Abbo, thus, appears to integrate empirical observations to bolster the understanding of the weight-determining role played by the qualities of wetness and dryness.

> And why should it be surprising that this happens with opposites, namely heat and cold, when the same is experienced in moisture and dryness, which alternately correspond to them? For indeed, the more something dries out by heating, the lighter it becomes in weight, and undoubtedly it becomes heavier by moistening through cooling. This is observed in logs

> full of sap and those partially burned, which have one part lighter than the other. For a log, suddenly thrown into water, bobs up more quickly on the burned side. We also know that the bodies of living creatures, after the vital heat has been taken away, grow heavier due to an abundance of corrupted blood. Therefore, while women are warmer during intercourse, are nevertheless colder due to their abundance of moisture – and we have learned that women drowned in rivers, because of the reproach of their sex, float prone on the water for longer periods, while the bodies of men who have been drowned always float supine. Finally, what else makes female limbs soft to the touch, boys or eunuchs with hairless bodies, if not the overflowing excess of cold moisture? What sharpens their high-pitched voices, if not the same principle observed in hydraulic organs? Indeed, for the same reason, we have found that elderly men get drunk more quickly than women, as they are relieved every month by the burden of the injurious [menstrual] flow. Some [women] do not conceive anything as we see in waterlogged places: the cast seeds would suffocate and become sterile due to excessive moisture. Thus, excessive moisture, like excessive dryness, leads to infertility, for if any one of the four qualities disproportionately prevails over the others, it causes those others to fail as if judged deficient, and thereby weight either increases or diminishes.[32]

Let us delve into the cases mentioned in the quotation above, as they provide insight into Abbo's unique approach of amalgamating literary sources with direct observations of natural phenomena to account for the action of the fundamental qualities in the physical realm. The first example is an fact that can be directly witnessed in nature, and notably absent in the literature of Abbo's time: a partially burned piece of wood exhibits distinct weight variations between the more and less burnt sections, correlating with the presence of lymph in the less burnt portion. This observation is further validated by the reaction observed when the burnt part of the ember is suddenly immersed in water, demonstrating that humidity increases the thickness of a body. The burnt part contracts rapidly, thickening and retracting towards the unburnt segment.

The second illustration associates more closely with the medical tradition, evident from the use of a specific term, namely, *sanies*, indicating some bodily discharge. Pliny employs this term in the *Historia naturalis* to denote the slime of the salamander and the humour of the viper. Additionally, he applies it in reference to the human body, generally signifying a liquid flowing from the ear.[33] Two other references to *sanies* appear more akin to Abbo's example. Celsus' definition, as presented in *De medicina*, *sanies* represents a weakened and corrupted form of blood, the pus discharged from an ulcer or wound.[34] The *Salernitan Questions* offer a slightly different definition, more related to the putrefactive state of a part of a body. To the question of why certain limbs or parts of some people become thinner or thicker than they should over time, the *Salernitan Questions* provide an answer that deals with the quality of heat: it is the overabundance of humour exposed to heat that can lead to putrefaction of the limb and subsequently, to the formation of *sanies* therein.[35] On his end, Abbo introduces the term *sanies* to characterise the specific post-mortem state of the body. The abundance of this liquid in the corpse – occurring once the living body has lost its vital heat – contributes to the body's increased weight. This explanation is also consistent with Macrobius' argument in his *Saturnalia*: as bodies are inherently subject to dissolution and putrefaction by nature, all bodies become heavier when the soul leaves them, transforming the once invigorating warm blood into corrupted blood (*sanies*).[36]

The *Saturnalia* seems to be the main source also for Abbo's discourse on moistness and wetness, with a focus on female physiology and the presence of liquids in their bodies. Characters in the *Saturnalia* repeatedly discuss women's moistness, emphasising two key aspects related to female fecundity: their ability to give birth (when the liquid is warm) and their menstrual cycle (when the liquid is cold). Oro, an Egyptian cynic, argues that the abundance of warm elements, like blood, in women makes them capable of generation.[37] Similarly, Flavianus, a prominent Roman politician, historian, and translator, asserts that warmth encourages love, stimulates seed ejaculation, and supports generation.[38] On the contrary, Quintus Aurelius Symmachus insists on the cold nature of women, advocating for 'frequent purges' (i.e., menstruation) due to an excess of unhealthy humour.[39] Disarius, whose medical theories are found throughout Book VII of the *Saturnalia*, defends the wet and cold nature of women. Among these theories on female physiology, Abbo appears to rely particularly on Oro's and Symmachus' perspectives. He contends that women, although warmer during intercourse, are inherently cold and moist, a characteristic demonstrated by menstrual flow. Beyond the physiological aspects, Abbo links this discourse about women to a moral notion derived from Pliny, who notes that the dead bodies of males float supine, those of females prone, as if nature acted out of a sense of decency.[40]

Furthermore, Abbo asserts that the excess of cold humours in women's bodies, as well as in those of children and eunuchs, results in smooth and hairless limbs. Macrobius, through Disarius, similarly explains that the smooth skin and 'monthly bleedings' in women are evidence of the moistness characterising the female body. For the same reason, childhood is considered a 'humid age' in human life, and eunuchs have notably delicate limbs.[41] Symmachus provides a more explicit statement on this matter:

> If women are not covered with hair like men, it is because of a lack of heat. Heat precisely makes hair grow: eunuchs lack it, and no one would deny that they have a colder nature than men.[42]

This explanation aligns with the *versio vetustissima* of the *Problemata physica*:

> Why eunuchs do not have beards? Answer: Because heat is blunted and wasted in them. For hair or fur grows from heat.[43]

The observation about the high-pitched voices of women, children, and eunuchs echoes a conversation between two characters in the *Saturnalia*, Disarius and Eusebius, the Greek master of rhetoric. In this exchange, Disarius is asked to explain why eunuchs have a high-pitched voice, making it challenging to distinguish them from women solely by their voices. Disarius attributes this characteristic to an excess of superfluous humour.[44] Abbo, possibly drawing from this dialogue, suggests that the high-pitched voices of women, children, and especially eunuchs are caused by the same element measured in hydraulic musical organs, which is water. Thus, what makes such voices so high-pitched is an excess of humours. Besides being an evidence of Abbo's interest in physiology, this reference also testifies to the spread and presence of a sophisticated musical instrument, such as the hydraulic organ, in Northern France.

Another observation Abbo makes about female physiology is related to alcohol consumption. He notes that women get drunk less quickly and less frequently than the elderly precisely due to the humours present in the female body. This account resonates with a similar explanation found once again in the *Saturnalia*. Flavianus asks Disarius to clarify Aristotle's statement from a supposed book titled *De ebrietate*, where Aristotle mentions that women rarely get drunk while the elderly do so often.[45] Disarius attributes this phenomenon to the abundance of humours in women, which dilutes the wine as it enters a body rich in liquids, preventing it from reaching the brain.[46]

Abbo's final remark regarding the quality of moistness is a unique synthesis of female physiology and botany. He draws a parallel between excessive moistness, which can render a woman sterile, and the inability of seeds scattered on the surface of spaded earth to grow due to excessive irrigation. This observation does not appear to be directly sourced from any specific text but reflects Abbo's own interpretation, possibly influenced by his understanding of both natural processes and the characteristics of female physiology.

Abbo's exploration of qualities also delves into a case study involving oil, honey, and wine – more precisely, their waste or dregs (*fex*). Oil and wine, appearing to be more humid than honey, expel their lees that deposit at the bottom of the container due to their greater density. In contrast, the lees of honey tend to sediment upwards. Considering this phenomenon, the purest parts of oil and wine are found respectively in the highest and middle parts of the containers. Conversely, the purest parts of honey are at the bottom. Abbo explains this by highlighting the reaction of liquids with air: being more humid than wine, oil reacts more favourably to the dry temperament of air, which balances its humidity. In half-empty wine containers, however, air dries the liquid excessively, making it too acidic. Furthermore, Abbo demonstrates the cold nature of wine by referring to its effect on the human body, specifically the tremors it induces, similar to those occurring in a feverish state. Regarding this effect, he discusses the physiological process at its core: wine, inherently cold, penetrates the veins and encounters the heat that accumulates especially in the region around the heart ("calor praecordialis").[47] The clash of these contrary qualities gives rise to a state of excitement and, consequently, the trembling of the body, resembling a febrile condition.

> The issue is about liquids, some of which are more moist, while others are more dry. But moist liquids, like wine and oil, send their sediment, which is denser and settles at the bottom, downwards. The sediment of drier substances, such as honey, on the other hand, springs upwards, as its purer drops, clinging to the bottom of the vessel, expel upwards the inferior dregs. Therefore, the best honey is believed to be at the bottom, the best wine in the middle, and the best oil at the top, because oil, being moister than wine, resists the tempering effect of air more easily. Thus, half-filled casks of wine are turned sour by the air drying them, while the liquid of oil, by the same process, becomes sweeter. Wine, being surrounded by air at the top and sediment at the bottom, ensures that its middle part remains unspoiled. Although it gladdens the heart of man, wine's nature is shown to be cold by the fact that it cause drunkards to shiver in a way familiar to those suffering from fever. For, as it penetrates the innermost veins, it encounters resistance from the heart's heat. This struggle introduces heat to the limbs, and the more they are filled with vigorous moisture, the less the cold from outside can exert its influence.[48]

Having clarified how fundamental qualities act in oil, wine, and honey, there comes Abbo's conclusive statement about their weight, which is as follows:

> And above all it must be known that the same quantity of oil or honey, received into one and the same vessel of equal capacity, differs in weights. And so, according to the hemiola ratio [i.e., the sesqualter, 3:2], the amount of honey outweighs an equal amount of oil by half of the latter. For example, if an eggshell holds an ounce of oil, it will also hold an ounce and a half of honey: indeed, within the same measure, the weights differ depending on the quality of the things.[49]

In this qualitative approach to physics, the issue of weight measurement comes into play, as substances of the same kind, such as lees of wine, oil, and honey, can behave differently based on their density and their interaction with the pure parts of the liquids they originate from. Abbo concedes that concerning weight measurement, when comparing equal volumes, liquids of the same amount exhibit specific weights: in fact, honey weighs one and a half times more than oil, which is nevertheless considered more humid.

Thus far, I have observed that Abbo's observations on physics and physiology seem to be a reinterpretation of certain arguments found in the *Saturnalia*.

This is especially evident concerning the behaviour of sediments in various liquid substances. To illustrate the extensive use of the *Saturnalia* by Abbo, I present below the examined passage from the *Explanatio* about sediments, comparing it in parallel with the sections of Book VII of the Saturnalia.

***In Calculo*, III, 93**	***Saturnalia*, VII[50]**	
But humid liquids, like wine and oil, send their sediment, which is denser and settles at the bottom, downwards. The sediment of drier substances, such as honey, on the other hand, springs upwards, as its purer drops, adhering to the bottom of the vessel, reject the inferior juice.	'Here is another query', said Avienus. 'Why do dregs always sink to the bottom, except the dregs of honey, which is unique in getting rid of its dregs by sending them to the top?' 'Dregs', replied Disarius, 'consist of a thick and earthy substance, and they are the heavier part of all liquids but honey, of which the dregs form the lighter part. In those other liquids, then, the dregs sink and fall to the bottom by reason of their weight, but in honey, since they are the lighter part, they give way and are forced from their place to the top'.	**§12, 8**
Therefore, honey is believed to be of the best quality at the bottom, wine in the middle, and oil at the top, because oil, being more humid than wine, more easily disregards the temperature of the air. Thus, by exposing wine in partially filled casks to the air, it turns into vinegar, while draining the liquid of oil makes it sweeter. Wine, being surrounded by air at the top and sediment at the bottom, ensures that its middle part remains uncorrupted.	'Hesiod', said Avienus (returning to the line of his inquiries), 'says that when you come to the middle of a jar of wine you should drink sparingly, although you may drink your fill of the rest of the jar, doubtless indicating thereby that the wine in the middle of a jar is the best. Experience too has proved that with oil the best part is that which floats at the top, and with honey the part which lies at the bottom. Why, pray, do we consider the oil at the top, the wine in the middle, and the honey at the bottom of the jar to be the best?' Without a moment's hesitation Disarius replied: 'With honey the best part is heavier than the rest, and so, in a jar of honey, since the part at the bottom is certainly the heaviest, it is therefore superior to that which floats above it. In a jar of wine, on the other hand, the lower part, being mixed with the dregs, is not only turbid but also inferior in flavour; and the wine at the top of the jar is spoiled by its proximity to the air, which mingles with the wine and weakens it'.	**§12, 13–14**

Although it cheers the heart of man, its cold nature is demonstrated by the fact that it induces feverish shivering in drunk people. As it penetrates the innermost veins, it encounters resistance from the heart's warmth. This struggle brings heat to the limbs, and the more robust the wine is with its liquid, the less the cold from outside can exert its influence.	But the clearest proof of the truth of my opinion is the fact that the effects of extreme cold and drunkenness are the same. In each condition the symptoms are shivering, heaviness, pallor, jerky and hurried breathing, and a trembling of the muscles and limbs; in each, the body is numb and the speech indistinct; and often, too, excess of wine, like exposure to extreme cold, results in the morbid state known to the Greeks as paralysis.	§6, 9

From this comparison, the direct influence of the dialogue between the characters Avienus and Disarius (*Saturnalia* VII, 12, 8, and 13–14) on the sediments of liquids and their reaction when in contact with air becomes evident as the primary source of Abbo's discussion. Additionally, Flavianus' remarks (*Saturnalia* VII, 6, 9) regarding the cold nature of wine can be regarded as the point from which Abbo elaborates on the effects of wine on the body. While Macrobius, through the character Flavianus, merely asserts a parallelism between the effects of drunkenness and cold – manifested as trembling, pallor, laboured breathing, difficulty in speech, and paralysis – Abbo goes a step further, providing a detailed explanation of the physiological processes occurring in the body in response to wine consumption. This illustrates how the physical and physiological arguments found in the *Saturnalia* have been synthesised, reorganised, and occasionally expanded by Abbo.

7.4 The 'natural power' of Things

In Abbo's perspective, qualitative physics seems to provide an explanation for a diverse range of phenomena. These phenomena span from cosmology to physiology, encompassing aspects such as the earth's position surrounded by water, air, and fire that compact it; the orbit and velocity of Saturn; the formation of ice under the Northern Circle; the properties of lead and its applications; the occurrence of lees in liquids; the behaviour of bodies in nature, particularly female bodies (their hairlessness, high-pitched voice, menstrual cycle, and infertility). All these phenomena are interpreted in the context of the influence of qualities, which play a crucial role in determining fundamental characteristics of substances in nature, including density, temperature, and consequently, weight. Within this framework, Abbo expands on the 'natural power' (*naturalis potentia*) of things. Such a 'natural power' operates through substantial quality. In this context, substantial quality may signify the inherent potential of a substance, that is, what a substance *has the power* to be. Abbo contends that it is through this natural power that further natural phenomena – even concerning mineralogy and metallurgy – can be elucidated, as we can read in the passage below.

> Therefore, natural power, which is the cause of heaviness or lightness in things, exerts itself through substantial quality. This is even experienced in the mass of quicksilver, which, while supporting the weight of a hundred stones, it immediately gives way when an ounce of gold is placed on it. To this one could also add the argument of the remarkable power of magnets, which suspend iron in the air. Whatever being, which thrives or thrived as animated by vital heat, tends to float on water, and most effectively can foods can be prepared by means of fire for living beings. Indeed, straw is believed to produce the best fire for shaping gold; fire from twigs is the healthiest for bodily heat; and the wood of the is the most suitable for melting glass. Thus, natural quality perfects the differences in things, such that one element agrees with another is some case and disagrees in others.[51]

Examples drawn from both the mineral and animal worlds are meant to further corroborate Abbo's idea that phenomena result from the presence of some natural power in things. The first case he illustrates is that of quicksilver, a metal previously mentioned by Pliny in the *Historia naturalis*, where it was noted that quicksilver corrodes any container and that all substances, except gold, float on it.[52] However, Abbo seems to derive the description of how quicksilver reacts to the weight of stones and gold from Isidore. According to Isidore, quicksilver can support the weight of a hundred stones but crumbles if pressed by a single *scripulus* of gold – the *scripulus* being equivalent to the 24th part of an ounce for Abbo.[53] This occurs due to the nature of gold itself causing quicksilver to collapse. A similar principle is observed with magnets, a phenomenon also described in Pliny's *Historia naturalis*, where he reports a mountain in Ethiopia that repels rather than attracts iron.[54] A phenomenon of this sort was also recounted by Isidore, who spoke of a temple where an iron statue was suspended in the void, along with the mountain in Ethiopia reported by Pliny.[55] In line with this records, Abbo mentions the case of magnetite, which, by repelling iron, makes it gravitate.

The last two examples involve living beings animated by vital heat: due to their fundamental quality, which is heat, each of them appears to be able to float in water, and for the same reason, their food is prepared using fire. All these instances aim to attest that substantial qualities reveal the 'natural power' inherent in things: metals and animals consistently respond to this, and their mutual reactions also demonstrate it. The substantial quality of a thing can contribute to the completion of another substance. Consider the example of fire, which Abbo draws once again from Macrobius' *Saturnalia*.[56] As Macrobius explains, there are different kinds of heat, and different properties depend on the 'bearer' of this quality: to forge gold, it is better to burn straw; to melt glass, the tamarisk tree is more suitable, while dry branches are preferable for health treatments using fire. What Abbo seems to convey is that, within a process of transforming substances, both a certain quality of the element used and a certain bearer of that element contribute to perfecting the substance undergoing transformation. This is enclosed within the notion of 'natural power' of things. To further grasp the concept of 'natural power' and its implementation through substantial qualities, one could also refer to Boethius' interpretation of Aristotle's category of quality. According to Aristotle, one type of quality is that of natural power – δύναμις φυσικὴ, translated precisely as *naturalis potentia* by Boethius – from which something derives its name. For example, a person is called a boxer not due to an inherent disposition but because they possess the physical power to easily engage in boxing.[57] Commenting on this type of quality, Boethius clarifies that individuals are labelled as boxers because of the sport discipline of boxing. Conversely, those who are not boxers but have the potential to become one are labelled as such not because of the sport discipline of boxing itself but because of the *power of this sport discipline*.[58] Therefore, according to Boethius, a quality can be acquired 'according to the natural power'. And it is against this philosophical backdrop that Abbo's concept of 'natural power' of things can be placed.

Notes

1 See Chenu, *La théologie*, 21–30; Gregory, "Considerazioni"; Gregory, "L'idea di natura", in which, however, the author considers Gerbert and Abbo as two distinctive exceptions in the 10th- and 11th-century scientific scenario (p. 85); Gregory, *La nouvelle*; Speer, *Die entdeckte Natur*.
2 Burnett, *Physics before the Physics*.
3 Germann, "Natural Philosophy", 223–227. On geometry and planetary astronomy between the 9th and 11th centuries, see also Eastwood, *The Revival*.
4 Caiazzo, "Filosofia della natura".
5 Caiazzo, "Filosofia della natura", 1079–1082.
6 For instance, the seventh chapter of Priscian's work draws from Aristotle's *De generation et corruption*, II, 3, focusing on "how elemental bodies undergo displacement up and down from their natural places". See, Priscian of Lydia, *Answers*, 59–69.
7 Beccaria, *I codici*, 152–156 and 193–197 respectively. The complete transcription of this codex has been made by Stoll, "Das Lorscher Arzneibuch".
8 Beccaria, *I codici*, 33.
9 On Fulbert's interest in medicine, see Fulbert of Chartres, *Epistulae*, 24, 47, and 49, pp. 44–46, 82 and 84–86. See also MacKinney, *Early Medieval Medicine*, 113 e 197, note 218; and MacKinney, *Bishop Fulbert*, 31–34.
10 Pseudo-Soranus, *Quaestiones medicinales*.
11 Caiazzo, "Filosofia della natura", 1081.
12 Lawn, *The Salernitan Questions*, 17–39 and 50–56.
13 Kristeller, *The School of Salerno*; Lawn, *The Salernitan Questions*, 16.
14 Richerius, *Historiarum libri*, II, 59, pp. 140–141. Cf. Lawn, *The Salernitan Questions*, 16.
15 Richerius, *Historiarum libri*, IV, 50, pp. 263–266.

16 Dan, "Les *Solutiones*". On the attribution of the *Solutiones ad Chosroem* to Eriugena, see D'Alverny, "Les *Solutiones*".
17 Ventura, "*Aristoteles fuit causa*", 113–117.
18 For his edition of the *versio vetustissima*, Valentin Rose collated three manuscripts, including the Bamberg one: Rose, *Aristoteles pseudepigraphus*, 666–676. For the transcription by Ulrich Stroll, see note 7 above. For the extant manuscripts of the *Problemata*, see Lawn, *The Salernitan Questions*, pp. 12–14.
19 As Macrobius himself states, his *Saturnalia* are inspired to the Platonic dialogues, and more precisely to Plato's *Symposium*. Macrobius, *Saturnalia*, I, 1, 3 and I, 1, 5, pp. 10 and 12, I vol.
20 I rely on Kaster, *Studies*.
21 Cassiodorus, *Expositio Psalmorum*, X, 7, p. 116: "Macrobius quoque Theodosius in quodam opere suo gentem dicit Cylicranorum fuisse iuxta Heracleam constitutam, composito nomen ἄπὸ τοῦ κύλιχος quod poculi genus una littera immutata calicem dixit". Cf. Macrobius, *Saturnalia*, V, 21, 18, p. 466, II vol: "Est etiam historia non adeo notissima nationem quandam hominum fuisse prope Heracleam ab Hercule constitutam Cylicranorum composito nomine ἀπὸ τῆς κύλικος, quod poculi genus nos una littera immutata calicem dicimus".
22 Arweiler, *Zu Text*.
23 Cuissard, *Catalogue*, vii-xviii; Delisle, *Notices*, 426–439.
24 For the link between the *Saturnalia* and Loup of Ferrières, see Marshall, "Macrobius", 222–235.
25 *In Calculo*, III, 91, p. 126: "Cur res eiusdem generis sunt graviores aliae aliis? Et ratio quidem in promptu est, maiorum subnixa institutis, quandoquidem quattuor elementorum diversitas ex quibus constant quattuor notissimis qualitatibus concordat. Sunt autem frigus et calor, humor et siccitas, quae, cum altrinsecus coniunguntur, nunquam se contraria herere patiuntur, licet ipsius naturae beneficio, quae levitati sunt obnoxia, omni nisu a se repellunt graviora".
26 Plinius the Elder, *Historia naturalis*, II, 10, pp. 10–11.
27 *In Calculo*, III, 91, p. 126: "Tanto enim unumquodque levius constat, quanto essentia caloris participat, tantoque fit gravius, quanto frigidius. Quis certe ignorat calorem ignis et aeris conprimere frigus aquae seu terrenae molis; terram scilicet hinc inde aequaliter libratam ab omni parte caeli, quod eam per singulos dies ex integro ambit, aquam vero ut corpulentiorem suis sufficere temperamentis?".
28 Aristotle, *Meteorology*, IV, 8.
29 Cf. Plinius the Elder, *Historia naturalis*, XXXIV, 166, p. 163: "In medicina per se plumbi usus cicatrices reprimere adalligatisque lumborum et renium parti lamnis frigidiore natura inhibere inpetus veneris".
30 Cf. Plinius the Elder, *Historia naturalis*, II, 6, 32–36, pp. 17–18; and Bede, *De natura rerum*, XIII, p. 204.
31 *In Calculo*, III, 91, pp. 126–127: "Et certe quicquid gelu stringitur, in se ipso densatur, tantoque fit densius quanto a calore remotius. Unde sub septentrionalis circulo glacialis aqua in lapidem vertitur, qui cristalus vocatur, nunquamque calore solis sursum hauritur refundenda, nisi congruo sit tepore marcida. Quid ergo mirum si quaedam terra frigore densatur, densata aliis gravior efficitur, cum quaedam aqua perpetuo rigore deorsum prematur? Quod enim plumbum ceteris metallis gravius naturaliter frigeat, adhletae noverunt, qui ob restringendam libidinem suis renis inponere consuerunt. Et Saturnus, altissimus planetarum, non solum itineris longitudine sed etiam algoris magnitudine pigrior vix duobus annis et semisse integris peragit XII partem zodiaci".
32 *In Calculo*, III, 92, pp. 127–128: "Et quid mirum, si id in contrariis, calore videlicet ac frigore, agitur, cum in humore ac siccitate, quae eisdem alternatim conveniunt, idem experiatur? Quanto enim unumquodque calendo siccatur, tanto pondere levius efficitur, quantoque frigendo humidius, tanto procul dubio gravius.

Quod animadvertitur in suci plenis et semiustis torribus, quorum pars altera levior habetur. Torris quippe, de repente iactu inmersus undis, ex ambusta parte citius resilit. Animantium quoque corpora, post subtractum vitae calorem, sanie abundantia novimus esse graviora. Quapropter, etsi ad coitum ferventiores, tamen humorum copia frigidiores feminas, flumine necatas, accepimus propter obprobium sexus superferri aquis pronas diutius, necatorum vero corpora semper natare resupina. Denique quid alius facit femineos artus tactu lenes, pueros vel eunuchos imberbes, nisi frigidi humoris inundans superfluitas? Quae causa eisdem tinnulas voces exacuit, nisi quae perpenditur in ydrauliis? Siquidem propter eandem rem senes cicius, mulieres tardius inebriari conperimus, quae singulis lunationibus sarcina noxii fluxus allevientur. Alias nil conciperent, quoniam ut in subsicivis locis videmus, iacta semina aquae praefocarent et sterilia nimio fluxu existerent. Sic nimirum nimius humor sicut nimia siccitas infecunditatem parit, quoniam una quattuor qualitatum, si reliquis inmutua varietate praeponderat, easdem ac si praeiudicium passas damnat, pondusque vel augmentat vel attenuat".

33 Plinius the Elder, *Historia naturalis*, X, 188; XI, 279; XXVII, 7; XXVIII, 94.
34 Celsus, *De medicina*, V, 26, 20B, p. 76, vol. 2.
35 Lawn, *The Prose*, B 120, pp. 56–57.
36 Macrobius, *Saturnalia*, VII, 12, 3, pp. 244–246.
37 Macrobius, *Saturnalia*, VII, 7, 4, pp. 210–121.
38 Macrobius, *Saturnalia*, VII, 6, 8, p. 204.
39 Macrobius, *Saturnalia*, VII, 7, 8–9, p. 214.
40 Plinius the Elder, *Historia naturalis*, VII, 77, p. 66: "Virorum cadavera supina fluitare, feminarum prona, velut pudori defunctarum parcente natura".
41 Macrobius, *Saturnalia*, VII, 6, 17, p. 208; Macrobius, *Saturnalia*, VII, 10, 8–11, pp. 238–240.
42 Macrobius *Saturnalia*, VII, 7, 8, pp. 212–214: "Nam quod pilis ut viri non obsidentur [*scil.* mulieres], inopia caloris est. Calor est enim qui pilos creat, unde et eunuchis desunt, quorum naturam nullus negaverit frigidiorem viris".
43 *Problemata*, 4, in Rose, *Aristoteles pseudepigraphus*, 667: "Quare spadoni non habent barbas. Resp. Quia obtunditur et deperit in eis calor ex calore enim crescunt capilli vel pili".
44 Macrobius, *Saturnalia*, VII, 10, 12–13, p. 240–241.
45 Cf. fr. 6 (94) in Rose, *Aristoteles pseudepigraphus*, pp. 118–119.
46 Macrobius, *Saturnalia*., VII, 6, 15–18, pp. 206–208. For a survey of the medieval practice of drinking wine, see Jaboulet-Vercherre, *The Physician*, 155–157 for the remarks concerning women.
47 On the *calor praecordialis* cf. Isidore, *Etymologiae*, XI, 1, 119, p. 78: "Praecordia sunt loca cordis vicina quibus sensus percipitur; et dicta praecordia eo quod ibi sit principium cordis et cogitationis".
48 *In Calculo*, III, 93, p. 128: "Argumento sunt liquida, quorum quaedam sunt humidiora, quaedam arida. Sed humida, ut vinum et oleum, feces utpote spissas deorsum subsidentes mittunt, <feces> aridorum vero, ut mellis, sursum resiliunt, quoniam eius puriores guttae fundo vasis herendo sursum despuunt obnoxia deteriori suco. Unde mellis creduntur optima quae sunt ima, vini quae media, olei quae summa, quia et oleum vino humidius contempnit aeris temperiem facilius. Quapropter vini semiplena dolia aer excoquendo in acorem vertit, olei vero liquorem exhauriendo dulcius efficit. Vinum quippe superius aere, inferius fece circumdatur, ut eius media incorrupta serventur, quod etsi cor hominis exhilarat, frigidum tamen natura illud demonstrat quod more febricitantibus familiari ebrios tremere cogit. Nam dum intima venarum penetrat, ei praecordialis calor repugnat. Quae pugna fervorem membris invehit, quae quanto virent sucis praevalida, tanto minus frigus a foris veniens penetrat suam exercet potentiam".

49 *In Calculo*, III, 94, pp. 128–129: "Sciendumque inprimis, quod eadem olei aut mellis quantitas, uno eodemque vase aequaliter recepta, pondere est diversa. Siquidem ratione hemiolii tota quantitas mellis propensior est ipsa medietate conmensurati olei. Ut verbi gratia, si testa ovi capit unciam olei, capiet quoque unciam et dimidiam mellis: quippe in eadem mensura sunt diversa pondera pro rerum qualitate".
50 Macrobius, *Saturnalia*, VII, 12 and 6. Translation by Vaughal Devies, in Macrobius, *The Saturnalia*, 490–491 and 469, respectively.
51 *In Calculo*, III, 91, p. 127: "Quocirca per substantialem qualitatem sic se exercet naturalis potentia, quae rebus est gravitatis aut levitatis causa, ut etiam experitur argenti vivi massa, quae cum sustentet molem centenarii lapidis, <molem> unciae auri superpositi ilico dehiscit. Accedit argumento magnetes mirae virtutis, qui ferrum in aere suspendit. Quicquid etiam animatum vitali calore viget aut viguit, aquis innatare consuevit et prout quidque animatorum valet maxime parari ignis alimoniae. Siquidem ad formandum aurum ignis creditur esse melior ex paleis; ad salutem corporis salubrior ex sarmentis; ad vitrum liquefaciendum habilior ex arbore cuius nomen est mirice. Sic naturalis qualitas perficit rerum differentias, ut quod elementum alii consenserat ab alio dissentiat".
52 Plinius the Elder, *Historia naturalis*, XXXIII, 99, p. 87.
53 Isidorus, *Etymologiae*, XVI, 19, 3, p. 215. For fractions and weights, included *scripuli*, see Chapter 6.
54 Plinius the Elder, *Historia naturalis*, XXXVI, 130, p. 95.
55 Isidorus, *Etymologiae*, XVI, 4, 2, p. 41.
56 Macrobius, *Saturnalia*, VII, 16, 22–23, p. 304; on gold, cf. Plinius the Elder, *Historia naturalis*, XXXIII, 60, p. 71.
57 *Boethius translator Aristotelis*, 8, pp. 24–25.
58 Boethius, *In Categorias*, PL 64, 253B-253C.

Conclusions

At the advent of the first millennium, a convergence of philosophical and scientific significance unfolded. Abbo of Fleury and his work, the *Explanatio in Calculo Victorii*, embodies this convergence and merit earnest consideration. The *Explanatio* stands out as pivotal contribution to the development of philosophical and mathematical thought in the Latin-speaking Europe. This contention has been substantiated first through an examination of Abbo's intellectual profile and key facets of his life: educational background, roles as a scholar and master in the liberal arts, his extensive travels and active involvement in institutional and political spheres. On account of these elements, he can be seen as a dynamic figure in the political and cultural landscape of medieval Europe.

In the history of mathematics and philosophy (and the history of their intertwinement), the *Explanatio* emerges as an original and systematic foray into Boethian number theory and at the same time serves as a comprehensive compendium of early medieval practical arithmetic, encapsulating the calculus of integers and fractions through methods employing fingers and tables. It is not just a mathematical text, that is, a commentary Abbo crafted on Victorius of Aquitaine's *Calculus*. However, to comprehensively evaluate both the scientific and philosophical breadth of the *Explanatio*, a preliminary examination of the *Calculus* was imperative. Victorius' *Calculus* comprises multiplication tables and supplementary arithmetical material, prefaced by an introductory section revolving around the notion of unity. By scrutinising the extant manuscripts of the *Calculus*, it was possible to discern the principal channels through which this work may have entered Fleury Abbey: one French (embodied by the character of Lupus of Ferrières) and one Irish (following the path of Columbanus). The analysis of the contexts in which Victorius' *Calculus* was available ascertained that it aligned with the presumptions underpinning time reckoning for the ecclesiastical calendar.

After the scrutiny of the *Calculus*, attention was directed towards the manuscript tradition of Abbo's *Explanatio*. It has been argued that preceding Alison Peden's endeavour, earlier editions of Abbo's *Explanatio* offered only a limited and fragmented comprehension of the text, predominantly concentrating on (yet not thoroughly uncovering) its numerical reckoning elements.

DOI: 10.4324/9781032643472-9

The prior editions failed to encapsulate not only the strictly mathematical facets but also the overall meaning and philosophical profundity of Abbo's commentary. While acknowledging the merits of Peden's critical edition, certain reservations have been expressed, particularly regarding the chaptering of the commentary. Drawing from manuscript evidence, I have proposed a distinct textual partition, deviating from the arrangement presented by Peden in her critical edition. This alternative approach yielded a more intelligible structure for the *Explanatio*, delineated into five sections: an *accessus*, a theological premise, a commentary on Victorius' *Preface*, a commentary on Victorius' multiplication table, and a section on qualitative physics – the latter was prompted by a segment of the *Calculus*' supplementary material, encompassing equivalence tables of units of measure. While this alternative partitioning enhances the clarity of Abbo's work, its fully appreciation required another preliminary step, consisting in the consideration of Abbo's account of the role of the commentator – with a view to answering the question 'Why to comment on a mathematical text?'. Drawing parallels with 12th-century perspectives, Abbo portrays the act of commenting on a text as an unveiling of its genuine meaning inherent within. His approach foreshadows notions such as *integumentum* and *involucrum* dear to the School of Chartres: Abbo presents the commentator, including himself, not merely as an explicator but as a revealer of authentic meanings concealed beneath obscure (philosophically charged) expressions. This unveiling is achieved through various means, including analogies, fictions, and myths, reflecting a distinctive approach to commentary. Thus, one could respond to the inquiry about the relevance of commenting on a mathematical text by anticipating that mathematics holds profound meanings that extend into the realm of philosophy (as well as theology and physics).

The analysis of Abbo's thought has been focused on the central, binomial concepts put forth in the *Explanatio*, specifically unity and composition. Prior to delving into these fundamental notions, a preliminary examination was conducted to elucidate the significance of what I called 'theological premise', a section within Abbo's *Explanatio* identified by the title *Tractatus de numero, mensura et pondere*. Indeed, as the title suggests, Abbo's *Tractatus* is situated within a longstanding exegetical tradition regarding *Wisdom* 11:21. Abbo's interpretation of this biblical verse unfolds along a twofold trajectory. On the one hand, there is the Augustinian perspective, which finds representation in Claudianus Mamertus' *De statu animae*. On the other hand, there is the mathematical reading, which is rooted in arithmetical assumptions derived from Boethius' *De arithmetica*, as exemplified in the works of John Scotus Eriugena and Hrabanus Maurus. These exegetical traditions merge in Abbo's *Tractatus* and *Explanatio*. Number, measure, and weight are exemplary causes according to which the world is created and structured. They are the conditions for beings to be quantified, namely, counted, measured, and weighed. Speculative arithmetic and calculus play a crucial role in understanding these foundational causes; for instance, by engaging in the study of speculative arithmetic and calculus, one can trace and comprehend the harmonious structure and organisation of created beings. The relationship between arithmetic and these exemplary causes also allows for unveiling God's mathematical rationality in the creation of the world. Thus, a 'theological premise' is essential to a mathematical text insofar as it points out the ultimate usefulness of such a scientific inquiry: understanding the divine (mathematical) causes inherent in the fabric of the universe.

The relevance of mathematics in conceptualising divine causes and the nature of created reality can be framed within Abbo's unequivocal Neopythagoreanism, which is also evident in his henology. Abbo's theory of unity has clear Neopythagorean influences that are primarily channelled through Boethius' writings. In this framework, Abbo draws parallels between God and the concept of arithmetic unity. This involves not only associating God with the notion of unity but also bringing the analogy of the relationship between unity and numbers into the picture. From this perspective, created beings and goods 'flow' from God, the highest good, like numbers and arithmetical and musical ratios 'flow' from unity and unison. This Neopythagorean standing is visually represented by Abbo's lambda-diagram, based on Calcidius and Macrobius, who introduced it to make sense of the generation of the World Soul recounted in Plato's *Timaeus*.

The conceptualisation of God as *individuum* originally adds to Abbo's Neopythagoreanism – the conceptualisation being rooted in Abbo's engagement with Boethius' logic writings. In this context, Abbo somehow appears to anticipate the 'universal singularism' that will be theorised in the 12th century. For according to Abbo, forms are viewed as *dividua*, representing common properties found in individual things. The collection of these *dividua* by likeness precisely gives rise to universals. This Aristotelian epistemological approach allows Abbo to conceive of God as *individuum*: in this perspective, God, who does not necessitate any formal determination of *dividua*, can be portrayed as truly one.

Abbo's thought extends the significance of unity beyond its application to God, recognising its key role in the ontology of all created beings. In Abbo's view, the fact that something exists is inseparable from the fact that it is one, and vice versa. While expanding on the parallelism of being and being one still on Boethian ground, Abbo can be seen as a precursor of some later debates on the convertibility of transcendental terms like 'one' and 'being'. Unity, then, is instrumental in conceiving the individual existence of created beings. Beyond safeguarding the individual existence of creatures, the concept of unity also allows for the conceptualisation of each being as a whole. In this respect, unity becomes a subject of interest in arithmetic and calculus. Through arithmetic and calculus, each creature is susceptible to disassembly, a process that involves breaking down and analysing the constituent elements of all things, corporeal and incorporeal.

The breakdown of beings considered as wholes is contingent upon a distinctive attribute inherent in creatures: composition. Indeed, all beings are composed, thereby distinguishing themselves from God, who is absolutely simple. Composition applies to both natural and artificial things, encompassing those compounds that do not depend on human will, as well as those resulting from it. Examples of the former include all natural phenomena (e.g., the synodic month and the lifespan of humans), while arithmetical ratios represent the latter, as they are produced according to rules established by a discipline. The evolution of natural and artificial compounds follows two interconnected stages that demonstrate an inverse proportionality to each other: growth and degrowth, production and reduction. Degrowth and reduction, in essence, elucidate how both natural entities and those shaped by human skill revert to their original constituents. This process involves a breakdown that retraces the same developmental stages, marking a methodical pattern in the life cycles and unfolding of all these entities.

The breakdown of beings into constituent elements also involves identifying a specific ontological composition, such as that of *esse* and *id quod est*, intrinsic to every creature. Abbo illustrates this primary composition by referring to the state of the earth at creation (*Gen.* 1:2). Although the earth is described as 'incomposite' immediately after creation, Abbo contends that its lack of composition actually pertains to its undifferentiated state with other elements rather than signifying true simplicity. Moreover, the composition of *esse* and *id quod est* is conferred as soon as whatever being receives the *forma essendi* from God, that is, at the moment of creation. Notions evidently drawn from Boethius' *De hebdomadibus*, then, contribute to a more comprehensive theory of composition.

The capacity for breakdown is inherent in all beings, irrespective of whether they possess a body. Corporeity, always associated with quantity, inherently suggests the potential for division, even when the body in question is not visibly perceptible. Conversely, the absence of a body does not preclude the possibility of division. Even abstract entities, including temporal measures, can be disassembled precisely because they are quantified, demonstrating that quantifiability (as well as composition) is not restricted to entities with a tangible, physical presence.

Given such a philosophical outline, one might not anticipate encountering Aristotelian physics in the *Explanatio*. However, this assertion is challenged by an examination of Abbo's conception of natural phenomena, particularly through the analysis of elementary qualities and their impact on the density of bodies, and consequently, their weight. Abbo's approach to physics dispenses with the metaphysical assumptions and doctrines presented elsewhere in his work, adopting what could be described as a Peripatetic – sometimes even 'empirical' – attitude. This departure from his broader philosophical framework in the treatment of natural phenomena is made possible by Abbo's reliance on indirect sources of Aristotelian physics, such as Macrobius' *Saturnalia*, which conveys Pseudo-Aristotle's *Problemata physica*. This is significant as it testifies to the presence of Aristotelian physics before Aristotle's *Physics* (and other texts on natural philosophy) made its appearance in Latin-speaking Europe through translations from Arabic. Abbo's dependence on such sources reveals the importance of Aristotelian principles in shaping early medieval philosophical views on nature, even before the direct availability of Aristotle's natural works in Latin translations.

This singular characteristic alone – i.e., the reliance on Aristotelian qualitative physics – provides ample reason to consider Abbo's text as deserving of scholarly attention. Hopefully, this book has brought to light several key facets which render Abbo and his text a substantial subject for scholars in both the fields of premodern history of science and philosophy. Firstly, an intricate exploration of late antique Platonic sources stands out in the *Explanatio*. Abbo engages in the comprehensive reading and interpretation of these sources, notably the works of Calcidius and Macrobius. This occurs at various levels, e.g., when it comes to understand the numerical generation of the World Soul, the symbolic values of certain numbers, for astronomical issues. This engagement with late antique Platonic works ultimately discards the idea that the first systematic reflection on them occurred in the 12th century. Secondly, Abbo testifies to a medieval Neopythagoreanism. He integrates a Neopythagorean perspective into his work, emphasising the role of mathematics as a tool for comprehending created beings, divine causes, and God. Thirdly, in the *Explanatio,* some later medieval theories seem to be somewhat in incubation. More specifically, Abbo's work exhibits elements that foreshadow certain 12th-century doctrines, e.g., the notion of *dividua* and the logical convertibility of 'being' and 'one', the notion of *integumentum*, and the Neopythagorean parallelism between numbers and creatures (doctrines dear to Thierry of Chartres, for instance). This forward-looking dimension suggests the pertinence of Abbo's ideas in influencing subsequent developments in medieval philosophical thought. Fourthly, Abbo aligns with the Boethian philosophical lineage, not only embracing Boethius' speculative arithmetic but also incorporating broader elements of Boethius' philosophy, encompassing arithmetic, theology, and logic. This contribution marks the beginning of a consolidation of Boethian philosophical elements that will become more pronounced precisely in the 12th century, the *aetas boethiana*. Lastly, Abbo's noteworthy contribution to the development of mathematics is evident. He emerges as one of the few premodern figures adept at synthesising speculative and practical arithmetic and combining operations with ten-based Roman numerals and twelve-based fractions.

Bibliography

Arithm. = Boethius, *De arithmetica*
Consolatio = Boethius, *Consolatio Philosophiae*
De statu animae = Mamertus, *De statu animae*
Hebd. = Boehtius, *De hebdomadibus*
In Calculo = Abbo of Fleury, *Commentary on the Calculus*
In somn. = Macrobius, *Commentarii in Somnium Scipionis*
In Tim. = Calcidius, *In Platonis Timaeum commentarium*
Musica = Boethius, *De institutione musica*
PL = Patrologia Latina cursus completus. Edited by J.-P. Migne. Paris, 1844–1855.
Trin. = Boethius, *De Trinitate*

Primary Sources

Abbo of Fleury. *Carmen acrosticum (Otto valens)*, PL 139, 519–520.

Abbo of Fleury. *Collectio canonum*, PL 139, 471–508.

Abbo of Fleury. *Commentary on the Calculus of Victorius of Aquitaine*. Edited by Alison Peden. Oxford: Oxford University Press, 2003.

Abbo of Fleury. *Commentary on the Calculus of Victorius of Aquitaine, Preface* [=Bede. *De ratione calculi libellus*], PL 90, 677–680.

Abbo of Fleury. *De duplici signorum ortu vel occasu. De cursu septem planetarum*. Edited by Ron B. Thomson. "Further astronomical Material of Abbo of Fleury". *Mediaeval Studies* 50 (1988): 671–673.

Abbo of Fleury. *De hypoteticis syllogismis*. Edited by Franz Schupp. Leiden, New York, and Köln: Brill, 1997.

Abbo of Fleury. *Dulce ingenium musicae*. Edited by Michel Bernhardt. *Anonymi saeculi decimi vel undecimi tractatus de musica 'Dulce ingenium musicae'*. München: Verlag der Bayerischen Akademie der Wissenschaften, 1987.

Abbo of Fleury. *Dulce ingenium musicae*. Edited by Michel Huglo. "Le traité de musique d'Abbon de Fleury. Identification et analyse". In *Abbon, un abbé de l'an mil*, edited by Anne Dufour and Gillette Labory, 225–239. Turnhout: Brepols, 2008.

Abbo of Fleury. *In patris natique simul*. Edited by Peter S. Baker and Michael Lapidge. "More Acrostic Verse by Abbo of Fleury". *Journal of Medieval Latin*, 7 (1997): 1–27.

Abbo of Fleury. *Liber apologeticus, PL* 139, 461–472.

Abbo of Fleury. *Quaestiones grammaticales*. Edited by Anita Guerreau-Jalabert. Paris: Les Belles Lettres, 1982.

Abbo of Fleury. *Sententia Abbonis de ratione spere*; *De duplici orto signorum; De quinque circulis mundi*. Edited by Ron B. Thomson. "Two Astronomical Tractates of Abbo of Fleury". In *The Light of Nature. Essays in the History and Philosophy of Science presented to A. C. Crombie*, edited by John D. North and J. J. Roche, 113–133. Dordrecht, Boston, and Lancaster: Martinus Nijhoff Publishers, 1985.

Abbo of Fleury. *Syllogismorum categoricorum et hypotheticorum enodatio*. Edited by André Van de Vyver. In *Abbonis Floriacensis Opera Inedita*, edited by R. Raes. Bruges: De Tempel, 1966.

Abbo of Fleury. *Victorii Calculi pars altera ex commentariis Abbonis Floriacensis adumbrata*. Edited by Wilhelm Christ. "Uber das argumentum calculandi des Victorius und dessen Commentar". In *Sitzungsberichte der Bayerische Akademie der Wissenschaften, Philos.-Philol., Classe 1*, 100–152. München: Franz, 1863.

Aimo of Fleury. *Vita Sancti Abbonis*. Edited and translated by Robert-Henru Bautier and Gllette Labory. *L'Abbaye de Fleury en l'an Mil*. Paris: CNRS Éditions, 2004.

Alcuin of York. *Disputatio de vera philosophia (De grammatica)*, PL 101, 847–902.

Anonymi saeculi decimi vel undecimi tractatus de musica Dulce ingenium musicae. Edited by Michael Bernhard. Munich: Verlag der Bayerischen Akademie der Wissenschaften, 1987.

Aristotle. *Predicamenta*. In *Boethius translator Aristotelis Categoriae (uel Praedicamenta)*. Edited by Lorenzo Minio Paluello. Bruges and Paris: Desclée De Brouwer, 1961.

Augustine. *Confessiones*. Edited by Lucas Verheijen. Turnhout: Brepols, 1983.

Augustine. *Contra Faustum manichaeum*. Edited by Joseph Zycha. Wien: Österreichische Akademie der Wissenschaften, 1891.

Augustine. *De civitate Dei*. Edited by Bernhard Dombart and Alfons Kalb. Turnhout: Brepols, 1955.

Augustine. *De Genesi ad litteram*. Edited by Joseph Zycha. Wien: Österreichische Akademie der Wissenschaften, 1894.

Augustine. *De Genesi contra Manichaeos*. Edited by Dorothea Weber. Wien: Österreichische Akademie der Wissenschaften, 1998.

Augustine. *De immortalitate animae*. Edited by Wolfgang Hörmann. Wien: Österreichische Akademie der Wissenschaften, 1986.

Augustine. *De musica*. Edited by Martin Jacobsson. Berlin and Boston: de Gruyter, 2017.

Augustine. *De ordine*. Edited by William M. Green. Turnhout: Brepols, 1970.

Augustine. *De Trinitate*. Edited by William J. Mountain. Turnhout: Brepols, 1968.

Augustine. *La Genèse au sens litteral en douze livres (I–VII)*. Introduction, traduction et notes par Paul Agaësse and Aime Solignac. Paris: Desclée De Brouwer, 1972.

Augustine. *The Literal Meaning of Genesis*. Translated by Edmund Hill. In *The Works of Saint Augustine. A Translation for the 21th Century*, 46 vols. John E. Rotelle et al. (eds.). Vol XIII. New York: New City Press, 1991–2019.

Aulus Gellius. *Noctae Attices*. Edited by Peter K. Marshall. Oxford: Clarendon Press, 1968.

Bedae opera de temporibus. Edited by Charles W. Jones. Cambridge, MA: Mediaeval Academy of America, 1943.

Bede. *De natura rerum*. Edited by Charles W. Jones. Turnhout: Brepols, 1975.

Bede. *De temporum ratione*. Edited by Charles W. Jones. Turnhout: Brepols, 1977.

Bede. *Opera exegetica libri quatuor in principium Genesis*. Edited by Charles W. Jones. Turnhout: Brepols, 1967.

Bede. *The Reckoning of Time*. Translated with introduction, notes, and commentary by Faith Wallis. Liverpool: Liverpool University Press, 2004.

Bernardus Silvestris (?). *Commentum super sex libros Eneidos Virgilii*. Edited by Wilhelm Riedel, Greiswald: Abel, 1924.

Boethius. *Commentaria in Categorias Aristotelis*, PL 64, 159–294.

Boethius. *Commentarii in Ciceronis Topica*. PL 64, 1039–1174.

Boethius. *Commentarii in librum Aristotelis Perihermeneias*. Edited by Karl Meiser. Leipzig: Teubner, 1877.

Boethius. *Consolatio*. Edited by Claudio Moreschini. Munich and Leipzig: de Gruyter, 2005.
Boethius. *Contra Euthychen et Nestorium*. Edited by Claudio Moreschini. Munich and Leipzig: de Gruyter, 2005.
Boethius. *De arithmetica*. Edited and Translated by Jean-Yves Guillaumin. Paris: Les Belles Lettres, 1995.
Boethius. *De arithmetica*. Translated by Michael Masi. *Boethian Number Theory. A Translation of De institutione arithmetica*. Amsterdam: Rodopi, 1983.
Boethius. *De hebdomadibus*. Edited by Claudio Moreschini. Munich and Leipzig: de Gruyter, 2005.
Boethius. *De institutione musica*. Edited by Gottfried Friedlein. Leipzig: Teubner, 1867.
Boethius. *De Trinitate*. Edited by Claudio Moreschini. Munich and Leipzig: de Gruyter, 2005.
Boethius. *In Isagogen Porphirii commenta. Editio prima*. Edited by Georg Schepps and Samuel Brandt, 1–132. Wien and Leipzig: F. Tempsky G. Freytag, 1906.
Boethius. *In Isagogen Porphirii commenta. Editio secunda*. Edited by Georg Schepps and Samuel Brandt, 135–348. Wien and Leipzig: F. Tempsky G. Freytag, 1906.
Bradwardine, Thomas. *Treatise on the Ratios of Velocities in Motion*. Edited by Lamar Crosby. *Thomas of Bradwardine: his Tractatus de Proportionibus: its significance for the development of mathematical physics*. Madison: University of Wisconsin Press, 1955.
Byrhtferth. *Enchiridion*. Edited by ed. Peter S. Baker and Michael Lapidge. Oxford: Early English Texts Society, 1995.
Calcidius. *In Platonis Timaeum commentarium*. Edited by Jan H. Wasnizk. London: Warburg Institute, 1975.
Cassiodorus. *De anima*. Edited by James W. Halporn. In *Traditio*, 16 (1960): 39–109.
Cassiodorus. *Expositio Psalmorum*. Edited by Marc Adriaen. Turnhout: Brepols, 1958.
Cassiodorus. *Institutiones divinarum et saecularium litterarum*. Edited by Roger A. B. Mynors. Oxford: Clarendon Press, 1937.
Celsus. *De medicina*. Edited by Walter G. Spencer. 3 vols. Cambridge, MA: Harvard University Press, 1961.
Cicero, Marcus Tullius. *De inventione*. Edited by Guy Achard. Paris: Les Belles Lettres, 1994.
Cicero, Marcus Tullius. *De oratore*, Edited by Edmond Courbaud. Paris: Les Belles Lettres, 1922.
Cicero, Marcus Tullius. *Topica*. Edited by Henri Bornecque. Paris: Les Belles Lettres, 1960.
Columbanus, *Epistulae*. In *Sancti Columbani Opera*, edited by G. S. M. Walker. Dublin: Dublin Institute for Advanced Studies, 1957.
Concilia Galliae A. 511 – A. 695. Edited by Charles de Clercq. Turnhout: Brepols, 1963.
De ponderibus. Edited by Friedrich Hultsch. *Metrologicorum scriptorum reliquiae*. Leipzig: Teubner, 1866.
Excerpta isagogarum et categoriarum. Edited by Giulio d'Onofrio. Turnhout: Brepols, 1995.
Faustus of Riez. *Epistula (Quaeris a me)*. In Claudianus Mamertus, *De statu animae*. Edited by August Engelbrecht, 10–11. Wien: Gerold, 1885.
Fulbert of Chartres. *Epistulae*. In Frederick Behrends. *The Letters and Poems of Fulbert of Chartres*. Oxford: Oxford University Press, 1976.
Gennadius of Massilia. *De viris illustribus*. Edited by William Herdin. Leipzig: Teubner, 1879.

Gerbert of Aurillac. *Epistulae*. Edited and translated by Jean-Pierre Callu and Pierre Riché. *Correspondance*. Paris: Les Belles Lettres, 2008.
Gerberti, postea Silvestri II papae, Opera mathematica (972–1003). Edited by Nicolai Bubnov. Berlin: Friedlander & Sohn, 1899.
Heriger of Lobbes. *Ratio numerorum abaci*. In *Gerberti, postea Silvestri II papae, Opera mathematica (972–1003)*, 221–225. Edited by Nicolai Bubnov. Berlin: Friedlander & Sohn, 1899.
Horologium stellare monasticum. Edited by Giles Constable. *Consuetudines Benedictinae variae (Saec. XI–Saec. XIV)*. Sieburg: Schmitt, 1975.
Hrabanus Maurus. *Commentariorum in librum Sapientiae libri tres*. PL 109, 671–762.
Hrabanus Maurus. *De computo*. Edited by Wesley M. Stevens. Turnhout: Brepols, 1979.
Hincmar of Reims, Pseudo. *De ratione animae*, PL 125, 932–937.
Hrabanus Maurus. *Commentaria in Genesim*. PL 107, 439–670.
Iamblichus. *The Theology of Arithmetic*. Translated by Robin Waterfield, with a Foreword by Keith Critchlow. Grand Raphis: Kairos, 1988.
Iamblichus. *Theologoumena arithmeticae*. Edited by Vittorio De Falco. Leipzig: Teubner, 1922.
Isidore of Seville. *Etymologiae*. Edited by Wallace M. Lindsay. Oxford: Clarendon Press, 1911.
Isidore of Seville. *Etymologiae II*. Edited by Peter K. Marshall and Jacques Fontaine. Paris: Les Belles Lettres, 1983.
Isidore of Seville. *Etymologiae III*. Edited by Giovanni Gasparotto and Jean-Yves Guillaumin. Paris: Les Belles Lettres, 2009.
Isidore of Seville. *Etymologiae V*. Edited by Valeriano Yarza Urquiola and Francisco J. Andrés Santos. Paris: Les Belles Lettres, 2013.
Isidore, *Etymologiae XI*. Edited by Fabio Gasti. Paris: Les Belles Lettres, 2010.
Isidore of Seville. *Etymologiae XVI*. Edited by José Feáns Landeira. Paris: Les Belles Lettres, 2011.
John of Holywood. *Algorismus vulgaris*. Edited by J. O. Halliwell. *Rara mathematica*. London: Metcalf and Palmer, 1839, 1–26.
John Scotus Eriugena. *Annotationes in Marcianum*. Edited by Cora E. Lutz. Cambridge, MA: The Medieval Academy of America, 1939.
John Scotus Eriugena. *Commentary on Boethius' Opuscula Sacra*. In Rand, Edward K. *Johannes Scottus. I. Der Kommentar des Johannes Scottus zu den Opuscula Sacra des Boethius, II. Der Kommentar des Remigius von Auxerre zu den Opuscula Sacra des Boethius*. Munich: Beck'sche, 1906.
John Scotus Eriugena. *De praedestinatione liber*. Edited by Ernesto S. Mainoldi. Florence: SISMEL-Edizioni del Galluzzo, 2003.
John Scotus Eriugena. *Periphyseon*. Edited by Édouard Jeaneau. 5 vols. Turnhout: Brepols, 1996–2008.
Jordanus de Nemore. *De elementis arithmetice artis*. Edited by Hubertus L.L. Busard, Stuttgart: Steiner, 1991.
Lupus of Ferrières. *Epistulae*. Edited by Peter K. Marshall. Leipzig: Teubner, 1984.
Macrobius. *Commentarii in Somnum Scopionis*. Edited by Mireille Armisen-Marchetti. 2 vols. Paris: Les Belles Lettres, 2011.
Macrobius. *Commentary on the Dream of Scipio*. Translated by William H. Stahl. New York: Columbia University Press, repr., 1990.
Macrobius. *Saturnalia*. Edited by Robert A. Kaster. 2 vols. Cambridge, MA: Harvard University Press, 2011.
Macrobius. *The Saturnalia*. Translated by Percival Vaughal Devies. New York-London: Columbia University Press, 1969.
Mamertus, Claudianus. *De statu animae*. Edited by August Engelbrecht. Wien: Gerold, 1885.

Martianus Capella. *De nuptiis Philologiae et Mercurii*. Edited by James Willis. Leipzig: Teubner, 1983.

Martianus Capella. *De nuptiis Philologiae et Mercurii. VII*. Edited and Translated by Jean-Yves Guillaumin. Paris: Les Belles Lettres, 2003.

Maximus the Confessor. *Ambigua ad Iohannem. Iuxta Iohannis Scotti Eriugenae Latinam Interpretationem*. Edited by Édouard Jeauneau. Turnhout: Brepols, 1988.

Micon of Saint-Riquier. *Carmina*. Edited by Ludwig Traube. Berlin: MGH *Poetae latini aevi Carolini* III, 1896.

Nicomachus of Gerasa. *Introduction to Arithmetic*. Translated into English by Martin L. D'ooge with Studies in Greek Arithmetic by F. E. Robbins and E. C. Karpinski. New York: The Macmillan Company, 1926.

Nicomachus of Gerasa. *Introductionis arithmeticae libri duo*. Edited by Richard G. Hoche. Leipzig: Teubner, 1866.

Notker of Liège. *De superparticularibus*. In *Gerberti, postea Silvestri II papae, Opera mathematica (972–1003)*. Edited by Nicolai Bubnov. Berlin: Friedlander & Sohn, 1899.

Plinius the Elder. *Historia naturalis II*. Edited by Jean Beaujeu. Paris: Les Belles Lettres, 1950.

Plinius the Elder. *Historia naturalis VII*. Edited by Robert Schilling. Paris: Les Belles Lettres, 1977.

Plinius the Elder. *Historia naturalis XXXIII*. Edited by Hubert Zehnacker. Paris: Les Belles Lettres, 1983.

Plinius the Elder. *Historia naturalis XXXIV*. Edited by Henri Le Bonniec. Paris: Les Belles Lettres, 1953.

Plinius the Elder. *Historia naturalis XXXVI*. Edited by Jacques André. Paris: Les Belles Lettres, 1981.

Priscian of Lydia. *Answers to King Khosroes of Persia*. Translated by Pamela Huby, Sten Ebbesen, David Langslow et al. London: Bloomsbury, 2016.

Proclus. *In Platonis Timaeum commentaria*. Edited by Ernst Diehl, Leipzig: Teubner, 1903.

Pseudo-Dyonisius. *Divine Names* (Eriugena's Latin translation). PL, 122, 1023–1194.

Pseudo-Soranus. *Quaestiones medicinales*. Edited by Klaus-Dietrich Fischer. Cuenca: Ediciones de la Universidad de Castilla-La Mancha, 2017.

Ratramnus of Corbie. *De anima*. Edited by A. Wilmart. "L'opuscule inédit de Ratramne sur la nature de l'âme". *In Revue Bénédictine*, 43 (1931): 207–223.

Remigius of Auxerre. *Commentum in Martianum Capellam*. Edited by Cora E. Lutz, 2 vols. Leiden: Brill, 1962–1965.

Remigius of Auxerre. *Expositio super Genesim*. Edited by Burton V. N. Edwards. Turnhout: Brepols, 1999.

Remmius Favinus. *Carmen de ponderibus et mensuris*. Edited by Alexander Riese. *Anthologia Latina sive Poesis Latinae, Supplementum I/2*. Leipzig: Teubner, 1906, pp. 29–37.

Remmius Favinus. *Carmen de ponderibus et mensuris*. Edited by Klaus Geus. Oberhaid: Utopica, 2007.

Richerius. *Historiarum libri*. Edited by Hartmuth Hoffmann. Hannover: Hahnsche, 2000.

Robert of Chester. *Euclid's Elements*. Edited by Hubertus L.L. Busard and Menso Folkerts. *Robert of Chester's (?) Redaction of Euclid's Elements. The so-called Adelard II Version*, Basel: Springer, 1992.

Roger Bacon. *Communia mathematica*. Edited by Robert Steele. Oxford: Clarendon Press, 1940.

Romana computatio. Edited by Charles Jones. *Beda pseudepigrapha: Scientific Writings Falsely Attributed to Bede*. Ithaca: Cornell University Press, 1939.

Sidonius Apollinaris. *Epistolae*. Edited by Jeffrey Henderson. Cambridge, MA and London: LOEB, 1965.

Sigebert of Gembloux. *De scriptoris ecclesiasticis*. PL, 160, 547–590.
Theon of Smyrna. *Expositio Rerum Mathematicarum adLegendum Platonem Utilium*, Introduzione, Traduzione, Commento di Francesco M. Petrucci. Sankt Augustin: Academia Verlag, 2012.
Thierry of Amorbach. *Consuetudines Floriacenses antiquiores*. Edited and translated by Anselme Davril and Lin Donnat. *Le coutumier de Fleury par Thierry d'Amorbach*. In *L'abbaye de Fleury en l'an mil*. Paris: CNRS Éditions, 2004.
Vetus latina. Die Reste der altlateinischen Bibel nach Petrus Sabatier neu gesammelt und in Verbindung mit der Heidelberger Akademie der Wissenschaftlichen herausgegeben von der Erzabtei Beuron unter der Leitung von Roger Gryson, 27 vols. *Sapientia Salomonis*. Vol. XI/1. Edited by Walter Thiele. Freiburg: Herder, 1977.
Victorius of Aquitaine. *Calculus*. Edited by Alison Peden. *Commentary on the Calculus of Victorius of Aquitaine*. Oxford: Oxford University Press, 2003.
Victorius of Aquitaine. *Calculus*. Edited by Gottfried Friedlein. "Der *Calculus* des Victorius", *Zeithschrift für Mathematik und Physik*, 16 (1871): 42–79.
Victorius of Aquitaine. *Calculus*. Edited by Gottfried Friedlein. "Victorii Calculus ex codice Vaticano editus", *Bullettino di bibliografia e di storia delle scienze matematiche e fisiche*, 4 (1871): 443–463.
Victorius of Aquitaine. *Calculus*. Edited by Wilhelm Christ. "Uber das argumentum calculandi des Victorius und dessen Commentar". In *Sitzungsberichte der Bayerische Akademie der Wissenschaften, Philos.-Philol., Classe 1*, 100–152. München: Franz, 1863.
Victorius of Aquitaine. *Praefatio*. Edited by Édmond Martène and Ursin Durand. *Thesaurus novus anecdotorum*. 118–120. Paris, 1717.
Victorius of Aquitaine. *Praefatio Commentarii in Cyclum Victorii*. PL 139, 569–572.
William of Conches. *Dragmaticon philosophiae*. Edited by Italo Ronca. Turnhout: Brepols, 1997.
William of Conches. *Glosae super Platonem*. Edited by Édouard Jeauneau. Paris: Vrin, 1965.

Secondary Literature

Aertsen, Jan. *Medieval Philosophy as Transcendental Thought. From Philip the Chancellor to Francisco Suarez*. Leiden: Brill, 2012.
Aertsen, Jan. "Ontology and Henology in Medieval Philosophy (Thomas Aquinas, Master Eckhart and Berthol of Moosburg)". In *On Proclus and His Influence in Medieval Philosophy*, edited by Egbert P. Bos and Pieter Anne Meijer, 120–140. Leiden, New York, and Köln: Brill, 1992.
Aertsen, Jan. "*Transcendens-Transcendentalis*. The Genealogy of a Philosophical Term". In *L'élaboration du vocaboulaire philosophique au Moyen Âge*, edited by Jacqueline Hamesse and Carlos Steel, 241–255. Turnhout: Brepols, 2000.
Albertson, David. *Mathematical Theologies. Nicholas of Cusa and the Legacy of Thierry of Chartres*. Oxford: Oxford University Press, 2015.
Ambrosetti, Nadia. *L'eredità arabo-islamica nelle scienze e nelle arti del calcolo nell'Europa medievale*. Milan: Edizioni universitarie di lettere economia diritto, 2008.
Arweiler, Alexander. "Zu Text und Überlieferung einer gekürzten Fassung von Macrobius *Saturnalia* I, 12, 2–I, 15, 20". In *Zeitschrift für Papyrologie und Epigraphik*, 131 (2000): 45–57.
Baker, Peter S., and Lapidge, Michael. "More Acrostic Verse by Abbo of Fleury". *Journal of Medieval Latin*, 7 (1997): 1–27.
Bakhouche, Béatrice. "Lectures musicales de l'harmonie musicale de l'âme selon Platon (*Timée* 35b-36b): l'influence de Calcidius". In *Revue de musicologie*, 98/2 (2008): 339–362.

Balmer, Robert T. "The Operation of Sandclocks and Their Medieval Development". In *Technology and Culture*, 19/4 (1978): 615–632.
Barker-Benfield, Bruce C. "A Ninth-Century Manuscript from Fleury: Cato de senectute cum Macrobio". In *Medieval Learning and Literature. Essays presented to Richard William Hunt*, edited by Jonathan J. G. Alexander and Margaret Gibson, 145–165. Oxford: Clarendon Press, 1976.
Barker-Benfield, Bruce C. "Macrobius". In *Texts and Transmission: A Survey of the Latin Classics*, edited by Leighton D. Reynolds. Oxford: Oxford University Press, 1983.
Barker-Benfield, Bruce C. The Manuscripts of Macrobius' Commentary on the Dream of Scipio, PhD Dissertation, 2 vols. Oxford University, 1975.
Batany, Jean. "Abbon de Fleury et les théories des structures sociales vers l'an mil", in *Études Ligériennes d'Histoire et d'Archéologie Médiévales. Mémoires et exposés présentés à la Semaine d'études médiévales de Saint-Benoît-sur-Loire, 3–10 juillet 1969*, edited by René Louis, 9–18. Paris: Clavreuil, 1975.
Beaujouan, Guy. "Étude paléographique sur la « rotation » des chiffres et l'emploi des apices du Xe au XIIe siècle". *Revue d'histoire des sciences*, 1–4 (1948): 301–313.
Beaujouan, Guy. "L'enseignement du *quadrivium*". In *La scuola nell'Occidente latino dell'alto medioevo*, 639–723. 2 vols. Spoleto: CISAM, 1972.
Beccaria, Augusto. *I codici di medicina del periodo presalernitano (secoli IX, X e XI)*. Rome: Edizioni Storia e Letteratura, 1956.
Beierwaltes, Werner. "Augustins Interpretation von *Sapientia 11:21*". In *Revue des études augustiniennes*, 15 (1969): 51–61.
Beierwaltes, Werner. *Denken des Eines: Studien zur Neuplatonischen Philosophie und ihrer Wirkungsgeschichte*. Frankfurt: Vittorio Klostermann, 1985.
Bettetini, Maria. *La misura delle cose. Struttura e modelli dell'universo secondo Agostino d'Ippona*. Milan: Rusconi, 1989.
Bezner, Frank. Vela veritatis. *Hermeneutik, Wissen und Sprache in der Intellectual History des 12. Jahrhunderts*. Leiden: Brill, 2005.
Bischoff, Bernhard. *Manuscripts and Libraries in the Age of Charlemagne*. Cambridge: Cambridge University Press, 1994.
Bischoff, Bernhard. *Mittelalterliche Studien. Ausgewählte Aufsätze zur Schriftkunde und Literaturgeschichte*. Stuttgart: Hiersemann, 1981.
Bömer, Franz. *Der lateinische Neuplatonismus und Neupythagorismus bei Claudianus Mamertus in Sprache und Philosophie*. Leipzig: Universitäts-buchdruckerei gebr. Scheur, 1936.
Bonazzi, Mauro. "Eudorus of Alexandria and the 'Pythagorean' Pseudepigrapha". In *On Pythagoreanism*, edited by Gabriele Cornelli, Richard McKirahan, and Constantinos Macris, 385–401. Berlin and Boston: de Gruyter, 2013.
Bonazzi, Mauro. "Plotino e la tradizione pitagorica". In *Acme*, 53 (2000): 39–73.
Bonazzi, Mauro, Lévy, Carlos, and Steel, Carlos. Eds. *A Platonic Pythagoras: Platonism and Pythagoreanism in the Imperial Age*. Turnhout: Brepols, 2007.
Borrelli, Arianna. *Aspects of the Astrolabe: «architectonica ratio» in Tenth- and Eleventh Century Europe*. Stuttgart: Franz Steiner, 2008.
Borst, Arno. *Astrolab und Klosterreform an der Jahrtausendwende*. Heidelberg: Ackademie der Wissenschaften, 1989.
Borst, Arno. *Das Mittelalterliche Zahlenkampfspiel*. Heidelberg: C. Winter, 1986.
Brisson, Luc. *Le même et l'autre dans la structure ontologique du Timée de Platon. Un commentaire systématique du Timée de Platon*. Paris: Klincksieck, 1974.
Brown, Nancy Marie. *The Abacus and the Cross: The Story of the Pope Who Brought the Light of Science to the Dark Ages*. New York: Basic Books, 2010.
Burnett, Charles. "Abbon de Fleury, *abaci doctor*". In *Abbon de Fleury. Philosophie, science et comput autour de l'an mil. Actes des journées organisées par le Centre d'histoire des sciences et des philosophies arabes et médiévales, CNRS/EPHE/Université Paris 7*, edited by Barbara Obrist, 129–139. Paris and Villejuif: CNRS, 2006.

Burnett, Charles. "Indian Numerals in the Mediterranean Basin in the Twelfth Century, with Special Reference to the "Eastern Forms"". In *From China to Paris: 2000 Years' Transmission of Mathematical Ideas*, edited by Yvonne Dold-Samplonius, Joseph W. Dauben, Menso Folkerts, and Benno van Dalen, 237–288. Stuttgart: Steiner, 2002.

Burnett, Charles. "Learning to Write Numerals in the Middle Ages". In *Teaching, Writing, Learning to Write: Proceedings of the XVI Colloquium of the Comité International de Paléographie Latine*, edited by Pamela Robinson, 233–240. London: King's College, 2010.

Burnett, Charles. "Physics before the Physics: Early Translations from Arabic of Texts concerning Nature in Mss British Library, Additional 22719 and Cotton Galba E IV". In *Medioevo*, 27 (2002): 53–109.

Burnett, Charles. "The Abacus at Echternach in ca. 1000 A.D.". In *Sciamus*, 3 (2002): 91–108.

Caiazzo, Irene. "Abbon de Fleury et l'héritage platonicien". In *Abbon de Fleury. Philosophie, science et comput autour de l'an mil. Actes des journées organisées par le Centre d'histoire des sciences et des philosophies arabes et médiévales, CNRS/EPHE/Université Paris 7*, edited by Barbara Obrist, 11–41. Paris and Villejuif: CNRS, 2006.

Caiazzo, Irene. "Filosofia della natura e fisica elementare nell'alto medioevo". In *La conoscenza scientifica nell'alto medioevo. (Settimane di Studio della CISAM. Spoleto 25 aprile – 1 maggio 2019)*, 1059–1084. 2 vols. Spoleto: CISAM, 2020.

Caiazzo, Irene. "La forme et les qualités des éléments: lectures médiévales du *Timée*". In *Il Timeo. Esegesi greche, arabe, latine*, edited by Francesco Celia and Angela Ulacco, 307–345. Pisa: Pisa University Press, 2012.

Caiazzo, Irene. "Mains célèbres dans les marges des *Commentarii in Somnium Scipionis* de Macrobe". In *'Scientia in margine'. Étude sur les marginalia in les manuscrits scientifiques du Moyen Âge à la Renaissance*, edited by Danielle Jacquart and Charles Burnett, 171–189. Genève and Paris: Droz-Champion, 2005.

Caiazzo, Irene, Macris, Constantinos, and Robert, Aurélien. Eds. *Brill's companion to the reception of Pythagoras and Pythagoreanism in the Middle Ages and the Renaissance*. Leiden: Brill, 2022.

Cantelli Berarducci, Silvia. *Hrabani Mauri opera exegetica. Repertorium fontium.* Turnhout: Brepols, 2006.

Cappuyns, Maïeul. *Jean Scot Érigène, sa vie, son oeuvre, sa pensée.* Louvain and Paris: Abbaye de Mont-César, 1933.

Carruthers, Mary. *The Book of Memory. A Study of Memory in Medieval Culture.* Cambridge: Cambridge University Press, 1990.

Casey, John Patrick. "Boethius's Works on Logic in the Middle Ages". In *A Companion to Boethius in the Middle Ages*, edited by Noel H. Kaylor and Philip E. Phillips, 193–220. Leiden and Boston: Brill, 2012.

Charbonnel, Nicole and Iung, Jean-Enric. Eds. *Gerbert l'européen. Actes du colloque d'Aurillac (Aurillac, 4–7 juin 1996).* Aurillac: Gerbert, 1997.

Chenu, Marie-Dominique. "*Involucrum. Le mythe selon les théologiens médiévaux*". In *Archives d'histoire doctrinale et littéraire du Moyen Âge*, 22 (1956): 75–79.

Chenu, Marie-Dominique. *La théologie au XII*[e] *siècle.* Paris: Vrin, 1957.

Chenu, Marie-Dominique. "Platon à Citeaux". In *Archives d'histoire doctrinale et littéraire du Moyen Âge*, 21 (1955): 99–106.

Christ, Wilhelm. "Uber das argumentum calculandi des Victorius und dessen Commentar". In *Sitzungsberichte der Bayerische Akademie der Wissenschaften, Philos.-Philol., Classe 1*, 100–152. München: Franz, 1863.

Cordoliani, Alfred. "A propos du premier chapitre du *De temporum ratione* de Bède". In *Le Moyen Âge*, 54 (1948): 209–223.

Cordoliani, Alfred. "Abbon de Fleury, Hériger de Lobbes et Gerland de Besançon sur l'ère de l'incarnation de Denys le Petit". *Revue d'histoire ecclésiastique*, 44 (1949): 463–487.

Cornford, Francis M. *Plato's Cosmology. The Timaeus of Plato Translated with a Running Commentary*. New York: Liberal Arts Press, 1957.

Corning, Caitlin. *The Celtic and Roman Traditions. Conflict and Consensus in the Early Medieval Church*. New York: Palgrave, 2006.

Courcelle, Pierre. *Les lettres grecques en Occident: de Macrobe à Cassiodore*. Paris: E. de Boccard, 1943.

Cousin, Patrice. *Abbon de Fleury-sur-Loire. Un savant, un pasteur, un martyr à la fin du Xe siècle*. Paris: Lethielleux, 1954.

Crialesi, Clelia V. "Absolute Spatial Differences: Grosseteste's Reading of Aristotle's *on the Heavens*". *Revista Española de Filosofía Medieval*, 30/1 (2023): 107–126.

Crialesi, Clelia V. "Les « raisons » des mathématiques: aspects de la ratio dans la pensée mathématique médiévale". In *La raison au Moyen Âge*, edited by Dominique Poirel, 77–96. Paris: Vrin, 2023.

Crialesi, Clelia V. "Numbers in Medieval Speculative Arithmetic". In *Quid sit numerus. Theories and Conceptualizations of Numbers in the Middle Ages and the Renaissance*, edited by Crialesi, Clelia V., Carole Hofstetter, and Maria Sorokina. Turnhout: Brepols, forthcoming.

Crialesi, Clelia V. "The Status of Mathematics in Boethius. Remarks in the Light of His Commentaries on the Isagoge". In *The Sustainability of Thought. An Itinerary Through the History of Philosophy*, edited by Lorenzo Giovannetti, 95–124. Naples: Bibliopolis, 2020.

Crialesi, Clelia V. "Un approccio matematizzante nell'analisi della realtà naturale: Abbone di Fleury e l'*Explanatio in Calculo Victorii*". In *Rappresentazioni della natura nel medioevo*, edited by Grassi, Onorato and Catapano, Giovanni, 41–58. Florence: SISMEL-Edizioni del Galluzzo, 2019.

Cristiani, Marta. "L'espace de l'âme. La controverse sur la corporéité des ésprits, le *De statu animae* de Claudien Mamert et le *Periphyseon*". In *Eriugena: Studien zu seinen Quellen. Vorträge des III Internationalen Eriugena-Colloquiums, Freiburg im Breisgau, 27–30 August 1979*, edited by Werner Beierwaltes, 149–163. Heidelberg: Carl Winter, 1980.

Cuissard, Charles. *Catalogue général des manuscrits des bibliothèques publiques de France*. XII vol. Orléans and Paris: Plon, 1889.

d'Alverny, Marie-Thérèse. "La Sagesse et ses sept filles. Recherches sur les allégories de la Philosophie et des arts libéraux du IXe au XIIe siècle". In *Mélanges dédiés à la mémoire de Félix Grat*, 245–278. 2 vols. Paris: Pecqueur-Grat, 1946.

d'Alverny, Marie-Thérèse. "Les *Solutiones ad Chosroem* de Priscianus Lydus et Jean Scot". In *Jean Scot Érigène et l'histoire de la philosophie*, edited by René Roques, 145–160. Paris: CNRS Éditions, 1977.

d'Onofrio, Giulio. "Introduction" to *Excerpta Isagogarum et Categoriarum*. Edited by Giulio d'Onofrio. Turnhout: Brepols, 1999.

d'Onofrio, Giulio. "Giovanni Scoto e Boezio: tracce degli *Opuscula sacra* e della *Consolatio* nell'opera eriugeniana". In *Studi medievali. III serie*, 21/2 (1980): 707–752.

Dachowsky, Elisabeth. *First among Abbots. The Career of Abbo of Fleury*. Washington: The Catholic University of America Press, 2008.

Dan, Anca. "Les *Solutiones ad Chosroem* de Priscien de Lydie et les transferts de savoirs pendant l'Antiquité tardive et le Moyen Âge". In *Orbis disciplinae. Hommages en l'honneur de Patrick Gautier Dalché*, edited by Nathalie Bernoux, Anca Dan, and Georges Tollias, 557–606. Turnhout: Brepols, 2017.

Davril, Anselme. "Un moine de Fleury aux environs de l'an mil: Thierry dit d'Amorbach". In *Études Ligériennes d'Histoire et d'Archéologie médiévale, Mémoires et exposés présentés à la Semaine d'études médiévales de Saint- Benoît-sur-Loire du 3 au 10 juillet 1969*, edited by René Louis, 97–105. Paris: Clavreuil, 1975.

De Libera, Alain. *L'art des généralités. Théories de l'abstraction*. Paris: Vrin, 1999.

Del Forno, Davide. "The mathematical structure of the soul of the world (Timaeus 35b4-36b6)". In *Elenchos*, 26/1 (2005): 5–32.

Delisle, Léopold. "Notices sur plusieurs manuscrits de la bibliothèque d'Orléans". In *Notices et extraits des manuscrits de la Bibliothèque Nationale et autres bibliothèques*. XXXI/1 vol. Paris: Imprimerie nationale, 1883.

Delville, Jean-Pierre, Kupper, Jean-Louis, and Laffineur Crepin, Marylène. Eds. *Notger de Liège. L'an mil au coeur de l'Europe*. Liège: Editions du Perron, 2008.

Dengler-Schreiber, Karin. *Scriptorium und Bibliothek des Klosters Michelsberg im Bamberg*. Graz: Akademische Druck- und Verlagsanstalt, 1979.

Depreux, Philippe, Lebecq, Stéphane, Perrin, Michel, and Szerwiniack, Olivier. Eds. *Raban Maur et son temps*. Turnhout: Brepols, 2010.

Dictionnaire de théologie catholique. Ed. Alfred Macant, Eugène Mangenot, and Emile Amann. Vols. 15. Paris: Letouzey, 1909–1950.

Di Marco, Michele. *La polemica sull'anima tra Fausto di Riez e Claudiano Mamerto*. Rome: Institutum Patristicum Augustinianum, 1995.

Dillon, John D. *Middle Platonism*. 80 B.C. to A.D. 220. Ithaca, NY: Cornell University Press, 1977.

Dohrn van Rossum, Gerhard. *History of the Hour: Clocks and Modern Temporal Orders*. Chicago and London: University of Chicago Press, 1996.

Dronke, Peter. *Fabula. Explorations into the Uses of Myth in Medieval Platonism*. Brill: Leiden and Köln, 1974.

Duchez, Marie-Élisabeth. "Jean Scot Erigène premier lecteur du '*De institutione musica*' de Boèce?". In *W. Eriugena: Studien zu seinen Quellen. Vorträge des III Internationalen Eriugena-Colloquiums, Freiburg im Breisgau, 27–30 August 1979*, edited by Werner Beierwaltes, 165–187. Heidelberg: Carl Winter, 1980.

Eastwood, Bruce. "Manuscripts of Macrobius, *Commentarii in somnium Scipionis*, before 1500". In *Manuscripta*, 38/2 (1994): 138–155.

Eastwood, Bruce. *The Revival of Planetary Astronomy in Carolingian and Post-Carolingian Europe*. Aldershot: Ashgate, 2002.

Engelen, Eva-Maria. *Zeit, Zahl un Bild. Studien zur Verbindung von Philosophie und Wissenschaft bei Abbo von Fleury*. Berlin-New York: de Gruyter, 1993.

Engels, Joseph. "Origine, sens et survie du term boécien *secundum placitum*". In *Vivarium*, 1 (1963): 87–114.

Evans, Gillian. "*Difficillima et ardua*": *Theory and Practice in Treatises on the Abacus, 950–110*". *Journal of Medieval History*, 3 (1977): 21–38.

Evans, Gillian. "Fractions and Fraction-Symbols in Boethius' Musica". *Centaurus*, 26 (1982): 115–217.

Evans, Gillian. "From Abacus to Algorism: Theory and Practice in Medieval Arithmetic". *British Journal for the History of Science*, 10/2 (1977): 114–131.

Evans, Gillian. "Introductions to Boethius' *Arithmetica* of the Tenth to the Fourteenth Century". *History of Science*, 16 (1978): 22–41.

Evans, Gillian. "Schools and Scholars: The Study of the Abacus in English Schools. C. 980 – 1150". *The English Historical Review*, 94 (1979): 71–89.

Evans, Gillian. "The *Rithmomachia*: A Medieval Mathematical Teaching Aid". In *Janus*, 63 (1976): 257–273.

Evans, Gillian. "The *Saltus Gerberti*: the Problem of the Leap". In *Janus*, 67/4 (1980): 261–268.

Evans, Gillian and Peden, Alison M. "Natural Sciences and the Liberal Art in Abbo of Fleury's Commentary on the Calculus of Victorius of Aquitaine". *Viator*, 16 (1985): 109–127.

Feichtinger, Hans. "'Nothing Rash Must Be Said': Augustine on Pythagoras". In *American Catholic Philosophical Quarterly*, 89/2 (2015): 253–276.
Ferrari, Franco. "I commentari specialistici alle sezioni matematiche del *Timeo*", In *La filosofia in età imperiale. Le scuole e le tradizioni filosofiche (Atti del colloquio, Roma, 17–19 giugno 1999)*, Edited by Aldo Brancacci, 171–224. Naples: Bibliopolis, 2000.
Ferrari, Mirella. "Spigolature bobbiesi. In margine ai Codices Latini Antiquiores". *Italia medievale e umanistica*, 16 (1973): 1–14.
Ferrari, Franco. "Struttura e funzione dell'esegesi testuale nel medioplatonismo: il caso del *Timeo*". In *Athenaeum*, 89 (2001): 525–570.
Flamant, Jacques. *Macrobe et le néoplatonisme latin à la fin du IVe siècle.* Leiden: Brill, 1977.
Folkerts, Menso. "Frühe Darstellungen des Gerbertschen Abakus". In *Itinera mathematica: Studi in onore di Gino Arrighi per il suo 90o compleanno*, edited by Raffaella Franci, Paolo Pagli, and Laura Toti Rigatelli, 23–43. Siena: Centro studi sulla matematica medioevale, 1996.
Folkerts, Menso. "Names and Forms of the Numerals on the Abacus in the Gerbert Tradition". In Menso Folkerts, *Essays on Early Medieval Mathematics, essay* no. VI. Repr. London and New York: Routledge, 2003.
Folkerts, Menso. "Names and Forms of the Numerals on the Abacus in Gerbert Tradition". In Menso Folkerts. *Essays on Early Medieval Mathematics. The Latin Tradition*, 1–17 (article VI). Aldershot: Ashgate, 2003.
Fortin, Ernest L. *Christianisme et culture philosophique au cinquième siècle. La querelle de l'âme humaine en Occident.* Paris: Études augustiniennes, 1959.
Galonnier, Alain. "Axiomatique et théologie dans le *De Hebdomadibus* de Boèce". In *Langages et philosophie. Hommage à Jean Jolivet*, edited by Alain De Libera, Abdelali Elamrani-Jamal, and Alain Galonnier, 311–330. Paris: Vrin, 1997.
Germann, Nadja. "À la recherche de la structure du temps: Abbon de Fleury et le comput". In *Abbon, un abbé de l'an mil*, edited by Anne Dufour and Gillette Labory, 153–176. Turnhout: Brepols, 2008.
Germann, Nadja. *De temporum ratione. Quadrivium und Gotteserkenntnis am Beispiel Abbos von Fleury un Hermanns von Reichenau.* Leiden and Boston: Brill, 2006.
Germann, Nadja. "Natural Philosophy in Early Latin Thought". In *The Cambridge History of Medieval Philosophy*, edited by Robert Pasnau and Christina Van Dyke, 219–230. Cambridge: Cambridge University Press, 2010.
Gibson, Margaret T. "The Continuity of Learning, circa 850-circa 1050". *Viator* 6 (1975): 1–13.
Gibson, Margaret T. Ed. *Boethius. His Life, Thought and Influence.* Oxford: Clarendon Prcss, 1981.
Gilson, Étienne. *L'être et l'essence.* Paris: Vrin, 1948.
Gougaud, Louis. "Les relations de l'abbaye de Fleury-sur-Loire avec la Bretagne armoricaine et les îles Britanniques (Xe et XIe siècles)". *Mémoires de la Société d'Histoire et d'Archéologie de Bretagne*, 3 (1922): 3–30.
Gracia, Jorge J. E. *Introduction to the Problem of Individuation in the Early Middle Ages.* Munich and Wien: Philosophia Verlag, 1984.
Gregory, Tullio. "Considerazioni per una storia del pensiero scientifico altomedievale". In *Studi medievali III serie*, 59 (2018): 271–282.
Gregory, Tullio. "La nouvelle idée de Nature et de savoir scientifique au XIIe siècle". In *The Cultural Context of Medieval Learning. Proceedings of the First International Colloquium in Philosophy, Science and Theology in the Middle Ages (September 1973)*, edited by John E. Murdoch and Edith D. Sylla, 193–218. Dordrecht and Boston: D. Reidel, 1975.
Gregory, Tullio. "L'idea di natura nella filosofia medievale prima dell'ingresso della fisica di Aristotele. Il secolo XII". In Tullio Gregory. *Mundana sapientia. Forme di conoscenza nella cultura medievale*, 77–114. Rome: Edizioni Storia e Letteratura, 1992.

Guerreau-Jalabert, Anita. "Introduction" to Abbo of Fleury, *Quaestiones grammaticales*. Edited by Anita Guerreau-Jalabert. Paris: Les Belles Lettres, 1982.
Gugel, Klaus. *Welche erhaltenen mittelalterlichen Handschriften dürfen der Bibliothek des Klosters Fulda zugerechnet werden?*, 2 vols. Frankfurt: Knecht, 1995, vol. I.
Guillaumi, Jean-Yves. "Boèce traducteur de Nicomaque: gnomons et pythmènes dans l'*Institution arithmétique*". In *Boèce ou la chaîne des savoirs. Actes du Colloque International de la Fondation Singer-Polignac. Paris 8–12 Juin 1999*, edited by Alain Galonnier, 341–356. Louvain-Paris: Peeters, 2003.
Guillaumin, Jean-Yves. "Boèce traducteur de Nicomaque: *gnomons* et *pythmènes* dans the *Institution Arithmétique*". In *Boèce ou la chaîne des savoirs, Actes du Colloque International de la Fondation Singer-Polignac, Paris, 8–12 juin 1999*, edited by Alain Galonnier, 341–355. Louvain and Paris: Peeters, 2005.
Guillaumin, Jean-Yves. "Boethius's De institutione arithmetica and Its Influence on Posterity". In *A Companion to Boethius in the Middle Ages*, edited Noel H. Kaylor and Philip E. Phillips, 135–161. Leiden and Boston: Brill, 2012.
Guillaumin, Jean-Yves. "Introduction". In Boethius. *De arithmetica*. Edited and Translated by Jean-Yves Guillaumin. Paris: Les Belles Lettres, 1995.
Guillaumin, Jean-Yves. "Le status des mathématiques chez Boèce". *Revue des Études Anciennes*, 92 (1990): 121–126.
Guillaumin, Jean-Yves. "L'ordre des sciences du "quadrivium" et la proportion géométrique". In *Latomus*, 50/3 (1991): 691–697.
Guiu, Adrian. Ed. *A Companion to John Scottus Eriugena*. Leiden and Boston: Brill, 2020.
Gwara, Scott. "Three Acrostic Poems by Abbo of Fleury". *Journal of Medieval Latin*, 1 (1991): 227–231.
Hadot, Pierre. "La distinction de l'être et de l'étant dans le *De hebdomadibus* de Boèce". In *Die Metaphysik im Mittelalter. Ihr Ursprung und ihre Bedeutung*, edited by Paul Wilpert, 147–153. Berlin: de Gruyter, 1963.
Hadot, Pierre. "La préhistoire des genres littéraires philosophiques médiévaux dans l'antiquité". In *Les genres littéraires dans le sources théologiques et philosophiques médiévales: définition, critique et exploitation. Actes du Colloque International de Louvain-la-Neuve (25–27 mai 1981)*, 1–9. Turnhout: Brepols, 1982.
Hagen, Hermann. *Catalogus codicum Bernensium*. Bern: Haller, 1875.
Handschin, Jacques. *The Timaeus schale*. In *Music discipline*, 4 (1950): 3–42.
Harrison, Carol. "Measure, Number and Weight in St. Augustine's Aesthetics". In *Augustinianum*, 28 (1988): 591–602.
Hart, Cyril. *Learning and Culture in Late Anglo-Saxon England and the Influence of Ramsey Abbey on the Major English Monastic Schools: The New Curriculum in Monastic Schools*. 2 vols. Lewiston: Edwin Mellen Press, 2003.
Hicks, Andrew. *Composing the World. Harmony in Medieval Platonic Cosmos*. Oxford: Oxford University Press, 2017.
Hicks, Andrew. "Pythagoras and Pythagoreanism in Late Antiquity and the Middle Ages". In *A History of Pythagoreanism*, edited by Carl A. Huffman, 416–434. Cambridge: Cambridge University Press, 2016.
Hoenig, Christina. *Plato's Timaeus in the Latin Tradition*. Cambridge: Cambridge university Press, 2018.
Hoenig, Christina. "Calcidius on Cosmic Harmony". In *Music and Philosophy in the Roman Empire*, edited by Federico M. Petrucci and Francesco Pelosi, 262–285. Cambridge: Cambridge University Press, 2020.
Holtz, Louis. "L'humanisme de Loup de Ferrières". In *Gli umanesimi medievali. Atti del II Congresso dell'«Internationales Mittelateinerkomitee», Firenze, Certosa del Galluzzo, 11–15 settembre 1993*, edited by Claudio Leonardi, 201–213. Florence: SISMEL-Edizioni del Galluzzo, 1998.

Huglo, Michel. "Compte rendu". In *Revue bénédictine*, 115 (2005): 225–226.
Huglo, Michel. "D'Helisachar à Abbon de Fleury". In *Revue bénédictine*, 104 (1994): 204–230.
Huglo, Michel. "Gerbert, théoricien de la musique, vu de l'an mil". *Cahier de civilisation médiévale*, 170 (2000): 143–160.
Huglo, Michel. "La réception de Calcidius et des *Commentarii* de Macrobe à l'époque carolingienne". In *Scriptorium*, 44 (1990): 3–20.
Huglo, Michel. "Le traité de musique d'Abbon de Fleury. Identification et analyse". In *Abbon, un abbé de l'an mil*, edited by Anne Dufour and Gillette Labory, 225–239. Turnhout: Brepols, 2008.
Huglo, Michel. "L'étude des diagrammes d'harmonique de Calcidius au Moyen Âge". In *Revue de musicologie*, 91/2 (2005): 305–319.
Huglo, Michel. "Trois livres manuscrits présentés par Helisachar". In *Revue bénédictine*, 99 (1989): 272–285.
Hultsch, Friedrich. *Metrologicorum scriptorum reliquiae*. Leipzig: Teubner, 1866.
Ifrah, George. *Histoire universelle des chiffres*. Paris, 1981. Repr. Paris: Bouquins, 2003.
Jaboulet-Vercherre, Azélina. *The Physician, the Drinker, and the Drunk. Wine's Uses and Abuses in Late Medieval Natural Philosophy*. Turnhout: Brepols, 2014.
Janby, Lars F. "Christ and Pythagoras. Augustine's Early Philosophy of Numbers". In *Platonism and Christian Thought in Late Antiquity*, edited by Panagiotis G. Pavlos, Lars F. Janby, Eyjólfur K. Emilsson, and Torstein T. Tollefsen, 117–128. London: Routledge, 2019.
Jeauneau, Édouard. *Études Érigéniennes*. Paris: Études Augustiniennes, 1987.
Jeauneau, Édouard. "Fulbert, notre vénérable Socrate". In *Fulbert de Chartres. Précurseur de l'Europe Médiévale?*, edited by Michel Rouche, 19–32. Paris: Presses Paris Sorbonne, 2008.
Jeauneau, Édouard. "Gloses et commentaires des textes philosophiques (IXe-XIIe s.)". In *Les genres littéraires dans le sources théologiques et philosophiques médiévales: définition, critique et exploitation. Actes du Colloque International de Louvain-la-Neuve (25–27 mai 1981)*, 117–131. Turnhout: Brepols, 1982.
Jeauneau, Édouard. "Jean Scot et la métaphysique des nombres". In *W. Eriugena: Studien zu seinen Quellen. Vorträge des III Internationalen Eriugena-Colloquiums, Freiburg im Breisgau, 27–30 August 1979*, edited by Werner Beierwaltes, 126–141. Heidelberg: Carl Winter, 1980.
Jeauneau, Édouard. "L'usage de la notion d'integumentum à travers les gloses de Guillaume de Conches". In *Archives d'histoire doctrinale et littéraire du Moyen Âge*, 24 (1957): 35–100.
Jones, Charles W. "The 'lost' Sirmond manuscript of Bede's computus". *English Historical Review*, 52 (1937): 204–219.
Jones, Charles W. "The Victorian and Dionysiac Tables in the West". *Speculum*, 9 (1934): 408–421.
Jurado, Ramos and Ángel, Enrique. "Moderato de Gades: estado de la cuestion. Cronologia y forma de vida". In *Habis*, 34 (2003): 149–160.
Juste, David. "Neither Observation nor Astronomical Tables: An Alternative Way of Computing Planetary Longitudes in Early Middle Ages". In *Studies in the History of Exact Sciences in Honour of David Pingree*, edited by Charles Burnett, Jan P. Hogendijk, Kim Plofker, and Michio Yano, 181–222. Leiden and Boston: Brill, 2004.
Kaster, Robert A. *Studies on the Text of Macrobius's* Saturnalia. Oxford: Oxford University Press, 2010.
Kinkelin, Hermann. *Der Calculus des Victorius, in Verhandlungen der Naturforschenden Gesellschaft in Basel*. Basel: Schweighauseriesche Verlagbuchhandlung, 1868, 147–158.

Klibansky, Raymond. *The Continuity of Platonic Tradition during the Middle Ages, (together with) Plato's Parmenides in the Middle Ages and Renaissance*. Munich and Millwood, NY: Fraus International Publications, 1981.
Kristeller, Paul. "The School of Salerno: Its Development and its Contribution to the History of Learning". In *Bulletin of the History of Medicine*, 17/2 (1945): 138–194.
Krusch, Bruno. *Studien zur christlich-mittelalteriche Chronologie. Die Entstehung unserer heutigen Zeitrechnung, Abhandlungen der preussischen Akademie der Wissenschaften. Jahrgang 1937*, Philosophisch-historische Klasse 8. Berlin: Verlag der Akademie der Wissenschaften, 1938.
Kühnel, Bianca. "Carolingian Diagrams, Images of the Invisible". In *Seeing the Invisible in Late Antiquity and the Early Middle Ages*, edited by Giselle de Niel, Karl F. Morrison, and Marco Mostert, 359–389. Turnhout: Brepols, 2005.
La Bonnardière, Annie-Marie. *Biblia augustiniana, A. T., Livre de la Sagesse*. Turnhout: Brepols, 1970.
La Bonnardière, Anne-Marie. "Le livre de la Sagesse dans l'oeuvre de Saint Augustin". In *Revue des études augustiniennes*, 17 (1971): 171–175.
Lambert, David. *Mamertus Claudianus*. In *The Oxford Guide to the Historical Reception of Augustine*, edited by Karla Pollmann and Willemien Otten. Vol III. Oxford: Oxford University Press, 2013.
Lapidge, Michael. "Litanies of the Saints in Anglo-Saxon Manuscripts: A Preliminary List". In *Scriptorium* 40 (1986): 264–277.
Lawn, Brian. *The Prose Salernitan Questions, edited from a Bodleian manuscript (Auct. F.3.10) an anonymous collection dealing with science and medicine written by an Englishman c. 1200*. London: British Academy, 1979.
Lawn, Brian. *The Salernitan Questions. An Introduction to the History of Medieval and Renaissance Problem Literature*. Oxford: Clarendon Press, 1963.
Leclercq, Henri. *Saint-Benoît-sur-Loire. Les reliques, le monastère, l'église*. Paris: Letouzey et Ané, 1925.
Lejbowicz, Max. "Postérité médiévale de la distinction isidorienne *astrologia/astronomia*: Bède et le vocabulaire de la chronométrie". In *Documents pour l'histoire du vocabulaire scientifique*, 7 (1985): 1–41.
Leonardi, Claudio. "Aspects of Old Testament Interpretation in the Church from the Seventh to the Tenth Century". In *Hebrew Bible/Old Testament. The History of Interpretation. From the Beginning to the Middle Ages (until 1300)*, edited by Magne Sæbø, 180–195. Göttingen: Vandenhoeck & Ruprecht, 2000.
Lesieur, Thierry. *Science des nombres et spiritualité entre Danube et Meuse (IXe-XIIe siècle)*. Paris: Classiques Garnie, 2009.
Löwe, Elias A. *Codices latini antiquiores, Supplement*, 1734, Oxford: Clarendon Press, 1971.
Luhtala, anneli. "Early Medieval Commentary on Priscian's *Institutiones grammaticae*". In *Cahiers de l'Institut du Moyen-Âge grec et latin*, 71 (2000): 115–88.
MacKinney, Loren. *Bishop Fulbert and Education at the School of Chartes*. Notre Dame (ind.): University of Notre Dame, 1957.
MacKinney, Loren. *Early Medieval Medicine. With Special Reference to France and Chartres*. Baltimore: Johns Hopkins Press, 1937.
Madan, Falconer. *A Summary Catalogue of Western Manuscripts in the Bodleian Library*. Oxford: Clarendon Press, 1895.
Maher, David W. and Makowsky, John F. "Literary Evidence for Roman Arithmetic with Fractions". In *Classical philology*, 96/4 (2001): 376–399.
Mainoldi, Ernesto S. "*Vox, sensus, intellectus* chez Jean Scot Érigène. Pour une focalisation des sources possibles du débat théologico-grammatical au XIe siècle". In *Arts du langage et théologie aux confins des XIe-XIIe siècles. Textes, maîtres, débats*, edited by Irène Rosier-Catach, 565–764. Turnhout: Brepols, 2011.

Marenbon, John. "Boèce, Porphyre et les variétés de l'abstractionnisme". In *Laval théologique et philosophique*, 68/1 (2012): 9–20.

Marenbon, John. *Early Medieval Philosophy 480–1150: An Introduction*. London and New York: Routledge, 1983.

Marenbon, John. *From the Circle of Alcuin to the School of Auxerre. Logic, Theology and Philosophy in the Early Middle Ages*. Cambridge: Cambridge University Press, 1981.

Marshall, Peter K. "Macrobius". In *Texts and Transmission. A Survey of the Latin Classics*, edited by Lighton D. Reynolds, 222–235. Oxford: Clarendon Press, 1983.

Masi, Michael. "The Influence of Boethius' *De arithmetica* on Late Medieval Mathematics". In *Boethius and the Liberal Arts. A Collection of Essays*, edited by Michael Masi, 81–96. Bern: P. Lang, 1981.

Materni, Marta. "Attività scientifiche di Gerberto d'Aurillac". In *Archivum Bobiense*, 29 (2007): 225–317.

Mathon, Gérard. *L'anthropologie chrétienne en Occident de saint Augustin à Jean Scot Erigène*. Lille: Université catholique de Lille, 1964.

McCarthy, Daniel. "The Arrival of the Latercus in Ireland". In *The Easter Controversy of Late Antiquity and the Early Middle Ages. Its Manuscripts, Texts and Tables, Proceedings of the 2nd International Conference on the Science of Computus in Ireland and Europe, Galway 18–20 July 2008*, edited by Immo Warntjes and Dáibhí Ó Cróinín, 48–75. Turnhout: Brepols, 2011.

McCarthy, Thomas J. H. *Music, Scholasticism and Reform. Salian Germany, 1024–1125*, Manchester: Manchester University Press, 2009.

McCluskey, Stephen C. *Astronomies and Cultures in Early Medieval Europe*. Cambridge: Cambridge University Press, 1998.

McCluskey, Stephen C. "Gregory of Tours, Monastic Timekeeping and Early Christian Attitudes to Astronomy". In *Isis*, 81 (1990): 8–22.

McDonald, Scott. "Boethius's Claim that all Substances are Good". In *Archiv fur Geschichte der Philosophie*, 70 (1988): 245–279.

Menninger, Karl. *Number Words and Number Symbols: A Cultural History of Number*. Cambridge, MA: MIT Press, 1969.

Mercati, Giovanni. "Le principali vicende del monastero di S. Colombano di Bobbio". In Giovanni Mercati, *Marci Tulli Ciceronis de re publica libri e codice rescripto Vaticano 5757 phototypice expressi*, 1–174. Città del Vaticano: Biblioteca Apostolica Vaticana, 1934.

Meyer, Christian. "L'âme du monde dans la rationalité musicale: ou l'expérience sensible d'un ordre intelligible". In *Harmonia mundi. Musica mondana e musica celeste fra Antichità e Medioevo*, edited by Marta Cristiani, Cecilia Panti, and Graziano Perillo, 57–75. Florence: SISMEL-Edizioni del Galluzzo, 2007.

Meyer, Christian. "La théorie des symphoniae selon Macrobe et sa diffusion". In *Scriptorium*, 53 (1999): 82–107.

Micaelli, Claudio. "Boethius's *De Hebdomadibus* in the Panorama of Late Ancient Thought". In *Boèce ou la chaîne des savoirs, Actes du Colloque International de la Fondation Singer-Polignac, Paris, 8–12 juin 1999*, edited by Alain Galonnier, 33–53. Louvain and Paris: Peeters, 2005.

Minois, J., "L'Horologium stellare monasticum a-t-il été écrit pour Fleury? Une approche géométrique et astronomique". *Abbon, un abbé de l'an mil*, edited by Anne Dufour and Gillette Labory, 47–68. Turnhout: Brepols, 2008.

Mostert, Marco. "Gerbert d'Aurillac, Abbon de Fleury et la culture de l'an mil: étude comparative de leurs oeuvres et de leur influence". In *Gerberto di Aurillac: da Abate di Bobbio a Papa dell'Anno 1000, Atti del congresso internazionale, Bobbio 28–30 settembre 2000*, edited by Nuvolone, Flavio G., 397–431. Bobbio: Archivum Bobiense, 2001.

Mostert, Marco. "La bibliothèque de Fleury-sur-Loire". In *Religion et culture autour de l'an mil. Royaume capétien et Lotharingie. Actes du colloque "Hugues Capet 987-1987. La France de l'an mil" (Auxerre, 26–27 juin 1987, Metz, 11–12 septembre 1987)*, edited by Dominique Iogna-Prat and Jean-Charles Picard, 119–123. Paris: Picard, 1990.

Mostert, Marco. *The Library of Fleury. A provisional List of Manuscripts*. Hilversum: Verloren, 1989.

Mostert, Marco. *The Political Theology of Abbo of Fleury. A Study of the Ideas about Society and Law of the Tenth Century Monastic Reform Movement*. Hilversum: Verloren Publishers, 1987.

Moyon, Marc, and Spiesser, Maryvonne. "L'arithmétique des fractions dans l'oeuvre de Fibonacci: fondements et usages". In *Archive for History of Exact Sciences*, 69 (2015): 391–427.

Murphy, James J. "Western Rhetoric in the Middle Ages". In *Latin Rhetoric and Education in the Middle Ages*, edited by James J. Murphy, 1–26. Ashgate: Aldershot, 2005.

North, John D. "Monasticism and the First Mechanical Clocks". In *The Study of Time II. Proceedings of the Second Conference of International Society for the Study of Time (Lake Yamanaka, Japan)*, edited by Julius T. Fraser and Nathaniel Lawrence, 381–398. New York, Heidelberg, and Berlin: Springer, 1975.

O'Meara, Dominic. *Pythagoras Revived. Mathematics and Philosophy in Late Antiquity*. Oxford: Oxford University Press, 1989.

O'Meara, Dominic. "The Metaphysical Use of Mathematical Concepts in Eriugena". In *Begriff und Metapher: Sprachform des Denkens bei Eriugena*, edited by. Werner Beyerwaltes, 142–148. Heidelberg: Carl Winter, 1990.

Obrist, Barbara. "Abbon de Fleury: cosmologie, comput, philosophie". In *Abbon, un abbé de l'an mil*, edited by Anne Dufour and Gillette Labory, 177–203. Turnhout: Brepols, 2008.

Obrist, Barbara. "Démontrer, montrer et l'évidence visuelle. Les figures cosmologiques de la fin de l'Antiquité à Guillaume de Conches et au début du XIIIe siècle". In *Diagramm und Text: Diagrammatische Strukturen und die Dynamisierung von Wissen und Erfharung*, edited by Eckart C. Lutz, Vera Jerjen, and Christine Putzo, 45–78. Wiesbaden: Reichert, 2014.

Obrist, Barbara. "Guillaume de Conches: cosmologie, physique du ciel et astronomie. Textes et images". In *Guillaume de Conches: philosophie et science au XIIe siècle*, edited by Barbara Obrist and Irene Caiazzo, 123–196. Florence: SISMEL-Edizioni del Galluzzo, 2011.

Obrist, Barbara. "Le tables et figures abboniennes dans l'histoire de l'iconographie des recueils de comput". In *Abbon de Fleury. Philosophie, science et comput autour de l'an mil (Actes des journées organisé par le Centre d'histoire des sciences et de philosophies arabes et médiévales)*, edited by Barbara Obrist, 141–186. Paris and Villejuif: CNRS, 2006.

Obrist, Barbara. "Wind, Diagrams and Medieval Cosmology". In *Speculum*, 72 (1997): 33–84.

Ohashi, Masako. "The Easter Table of Victorius of Aquitaine in Early Medieval England". In *The Easter Controversy of Late Antiquity and the Early Middle Ages. Its Manuscripts, Texts and Tables, Proceedings of the 2nd International Conference on the Science of Computus in Ireland and Europe, Galway 18–20 July 2008*, edited by Immo Warntjes and Dáibhí Ó Cróinín, 137–149. Turnhout: Brepols, 2011.

Otisk, Marek. *Arithmetic in the Thought of Gerbert of Aurillac*. Berlin: Peter Lang, 2022.

Otisk, Marek. "Letter on Timekeeping of Gerbert of Aurillac to Brother Adam". In *Konštantínove Listy*, 11/1 (2018): 67–78.

Otisk, Marek. "The Interpretations and Applications of Boethius's *Introduction to the Arithmetic* II, 1 at the End of the 10th Century". In *Gerbertus*, 2 (2011): 33–56.

Otten, Willemien and Allen, Michael. Eds. *Eriugena and Creation. Proceedings of the Eleventh International Conference on Eriugenian Studies, held in honor of Édouard Jeauneau, Chicago, 9–12 November 2011*. Turnhout: Brepols, 2014.
Panti, Cecilia. "Boethius and Ptolemy on Harmony, Harmonics and Human Music". In *Micrologus*, 25 (2017): 3–35.
Panti, Cecilia. *Filosofia della musica. Tarda antichità e medioevo*. Rome: Carocci, 2008.
Panti, Cecilia. "Pythagoras and the Quadrivium from Late Antiquity to the Middle Ages". In *Brill's Companion to the Reception of Pythagoras and Pythagoreanism in the Middle Ages and the Renaissance*, edited by Irene Caiazzo, Constantinos Macris, and Aurélien Robert, 47–81. Turnhout: Brepols, 2021.
Peden (White), Alison, M. "Boethius in the Medieval Quadrivium". In *Boethius. His Life, Thought and Influence*, edited by Gibson, Margaret, 162–205. Oxford: Blackwell, 1981.
Peden (White), Alison. *Glosses Composed before the Twelfth Century in Manuscripts of Macrobius' Commentary on Cicero's Somnium Scipionis*. 2 vols. Ph.D. diss., Oxford University, 1981.
Peden (White), Alison, M. "Introduction" to Abbo of Fleury, *Commentary on the Calculus of Victorius of Aquitaine*. Edited by Alison Peden. Oxford: Oxford University Press, 2003.
Peden (White), Alison, M. "Unity, Order and Ottonian Kingship in the Thought of Abbo of Fleury". In *Belief and Culture in the Middle Ages: Studies in the Medieval Church presented to Henry Mayr-Harting*, edited by Gameson, Richard and Leyser, Henrietta, 158–168. Oxford: Oxford University Press, 2001.
Pellegrin, Elisabeth. "La tradition des textes classiques latins à l'abbaye de Fleury-sur-Loire", *Revue d'histoire des textes* 14–15 (1984–1985): 155–167.
Pellegrin, Elisabeth. "Membra disiecta floriacensa". *Bibliothèque de l'École des chartes* 117 (1959): 5–56.
Pelosi, Francesco. *Plato on Music, Soul and Body*. Cambridge: Cambridge University Press, 2010.
Peri, Israel. "*Omnia mensura et numero et pondere disposuisti*: die Auslegung von Weish. XI, 20 in der lateinischen Patristik". In *Mass, Zahl, Zahlensymbolik im Mittelalter*, edited by Albert Zimmermann, 1–21. De Gruyter: Berlin-New York, 1983.
Pessin, Sarah. "Hebdomads: Boethius meets the Neopythagoreans". In *Journal of the History of Philosophy*, 37/1 (1999): 29–48.
Pizzani, Ubaldo. "Il *quadrivium* boeziano e i suoi problemi". In *Atti del congresso internazionale di studi boeziani (Pavia, 5–8 ottobre 1980)*, edited by Luca Obertello, 211–226. Pavia: Herder, 1981.
Pizzani, Ubaldo. "Qualche osservazione sul concetto di armonia cosmica in Agostino e Cassiodoro alla luce di Sap 11,21(20)". In *Augustinianum*, 32/2 (1992): 301–322.
Prou, Maurice and Vidier, Alexandre. *Recueil des chartes de l'abbaye de Saint-Benoît-sur-Loire*, 1–19. Paris: Société historique et archéologique du Gâtinais, 1900, vol. I.
Quain, Edwin A. "Medieval *accessus ad auctores*". In *Traditio*, 5 (1985): 215–264.
Rand, Edward K. *Johannes Scottus. I. Der Kommentar des Johannes Scottus zu den Opuscula Sacra des Boethius, II. Der Kommentar des Remigius von Auxerre zu den Opuscula Sacra des Boethius*. Munich: Beck'sche, 1906.
Restani, Donatella. "Le radici antropologiche dell'estetica boeziana. *Anima humana e musica humana*. In *Musica e storia*, 2 (2007): 243–258.
Reydams-Schils, Gretchen. *Calcidius on Plato's Timaeus*. Cambridge: Cambridge University Press, 2020.
Ricciardi, Alberto. *L'Epistolario di Lupo di Ferrières. Intellettuali, relazioni culturali e politica nell'età di Carlo il Calvo*. Spoleto: Fondazione Centro italiano di studi sull'alto Medioevo, 2005.
Riché, Pierre. *Abbon de Fleury. Un moine savant et combatif (vers 950–1004)*. Turnhout: Brepols, 2004.

Riché, Pierre. *Écoles et enseignement dans le Haut Moyen Âge: fin du Ve siècle-milieu du XIe siècle*. Paris: Picard, 1999.
Riché, Pierre. "Le *quadrivium* dans le Haut Moyen Âge". In *Scienze matematiche e insegnamento in epoca medioevale. Atti del convegno internazionale di studio, Chieti, 2–4 maggio 1996)*, edited by Luigi Pellegrini, Paolo Freguglia, and Roberto Paciocco, 93–111. Naples: Edizioni Scientifiche Italiane, 2000.
Riché, Pierre. *L'enseignement au Moyen Âge*. Paris: CNRS Éditions, 2016.
Riché, Pierre. "Relations entre l'abbaye de Fleury-sur-Loire et les pays celtiques (X^e^-XIe siècles)". In *Corona monastica. Moines bretons de Landévennec: histoire et mémoire celtiques. Mélanges offerts au père Marc Simon*, edited by Louis Lemoine and Bernard Merdignac, 13–18. Rennes: Presse Universitaire de Rennes, 2004, pp. 13–18.
Riché, Pierre and Verger, Jacques. *Des nains sur des épaules de géants. Maîtres et élèves au Moyen Âge*. Paris: Tallandier, 2006.
Richter, Michael. *Bobbio in the Early Middle Ages. The Abiding Legacy of Columbanus*. Dublin: Four Courts Press, 2008.
Robbins, Frank E. "The Tradition of Greek Arithmology". In *Classical Philology*, 16 (1921): 97–123.
Robbins, Frank E., and Karpinski, Louis C. "Studies in Greek Arithmetic". In Nicomachus of Gerasa. *Introduction to Arithmetic*. Translated into English by Martin L. D'ooge with Studies in Greek Arithmetic by F. E. Robbins and E. C. Karpinski, 3–177. New York: The Macmillan Company, 1926.
Roche, William J. "Measure, Number and Weight in saint Augustine". In *New Scholasticism*, 15 (1941): 350–376.
Romano, Antonio. "Lupo di Ferrières, un umanista del IX secolo". In *Gli umanesimi medievali. Atti del II Congresso dell'«Internationales Mittelateinerkomitee», Firenze, Certosa del Galluzzo, 11–15 settembre 1993*, edited by Claudio Leonardi, 583–589. Florence: SISMEL-Edizioni del Galluzzo, 1998.
Rose, Valentin. *Aristoteles pseudepigraphus*. Leipzig: Teubner, 1893.
Rose, Valentin. *Verzeichnis der lateinischen Handschriften der Königlichen Bibliothek zu Berlin*, vol. 1. Berlin: Georg Olms, 1893.
Rosier-Catach, Irène. "Regards croisés sur le pouvoir des mots au Moyen Âge". In *Le pouvoir des mots au Moyen Âge*, edited by Nicole Bériou, Jean-Patrice Boudet, and Irène Rosier-Catach, 511–585. Turnhout: Brepols, 2014.
Rosier-Catach, Irène. "*Vox* and *oratio* in Early Twelfth Century Grammar and Dialectics". In *Archives d'histoire doctrinale et littéraire du Moyen Âge*, 78 (2011): 47–129.
Savigni, Raffaele. "L'interpretazione dei libri sapienziali in Rabano Mauro: tradizione patristica e *moderna tempora*". In *Annali di storia dell'esegesi*, 9 (1992): 557–587.
Schupp, Franz. "Abbon de Fleury et la logique: quelques questions historiques et systematiques". In *Abbon de Fleury. Philosophie, science et comput autour de l'an mil. Actes des journées organisées par le Centre d'histoire des sciences et des philosophies arabes et médiévales, CNRS/EPHE/Université Paris 7*, edited by Barbara Obrist, 43–59. Paris and Villejuif: CNRS, 2006.
Solère, Jean-Luc. "Bien, cercles et hebdomades. Formes et raisonnement chez Boèce et Proclus". In *Boèce ou la chaîne des savoirs, Actes du Colloque International de la Fondation Singer-Polignac, Paris, 8–12 juin 1999*, edited by Alain Galonnier, 55–110. Louvain and Paris: Peeters, 2005.
Somfai, Anna. "Calcidius' Commentary on Plato's *Timaeus* and Its Place in Commentary Tradition: the Concept of Analogy in Text and Diagrams". In *Bulletin of the Institute of Classical Studies. Supplement*, 83 (2004): 203–220.
Somfai, Anna. "The Brussels Gloss: A Tenth-Century Reading of the Geometrical and Arithmetical Passages of Calcidius's Commentary (ca. 400 AD) to Plato's *Timaeus*". In *Scientia in margine. Études sur les marginalia dans les manuscrits scientifiques du Moyen Âge à la Renaissance*, edited by Danielle Jacquart and Charles Burnett, 139–169. Geneva and Paris: Droz-Champion, 2005.

Somfai, Anna. "The Eleventh-Century Shift in the Reception of Plato's *Timaeus* and Calcidius's *Commentary*". *Journal of the Warburg and Courtauld Institutes*, 65 (2002): 1–21.
Speer, Andreas. *Die entdeckte Natur. Untersuchungen zu Begründungsversuchen einer scientia naturalis im 12. Jahrundert.* Leiden: Brill, 1995.
Stahl, Irene. *Katalog der mittelalterlichen Handschriften der Staats- und Universitätsbibliothek Bremen.* Wiesbaden: Harrassowitz, 2004.
Stoll, Ulrich. *Das Lorscher Arzneibuch. Ein medizinisches Kompendium des 8. Jahrhunderts (Codex Bambergensis medicinalis 1). Text, Übersetzung und Fachglossar.* Stuttgart: Steiner 1992.
Suto, Taki. *Boethius on Mind, Grammar and Logic. A Study on Boethius' Commentaries on Peri Hermeneias.* Leiden and Boston: Brill, 2012.
Tabulae codicum manu scriptorum praeter graecos et orientales in Bibliotheca Palatina Vindobonensi asservatorum, 11 vols. Wien: Caroli Geroldi, 1868, vol. II.
Tannery, Paul. "Sur les subdivisions de l'heure durant l'Antiquité". *Revue d'archéologie*, 26/1 (1895): 359–366.
Thomson, Ron B. "Further astronomical Material of Abbo of Fleury", *Mediaeval Studies* 50 (1988): 671–673.
Thomson, Ron B. "Two Astronomical Tractates of Abbo of Fleury". In *The Light of Nature. Essays in the History and Philosophy of Science presented to A. C. Crombie*, edited by John D. North and J. J. Roche, 113–133. Dordrecht, Boston, and Lancaster: Martinus Nijhoff Publishers, 1985.
Tolomio, Ilario. *L'anima dell'uomo. Trattati sull'anima dal V al IX secolo.* Milan: Rusconi, 1979.
Tolomio, Ilario. "L'origine dell'anima nell'altro medioevo. La fortuna della tesi di san Girolamo: Dio crea l'anima di ogni uomo che viene in questo mondo". In *Medioevo*, 13 (1987): 51–73.
Tornau, Christian. "Intelligible Matter and the Genesis of Intellect: The Metamorphosis of a Plotinian Theme in *Confessions* 12–13". In *Augustine's Confessions: Philosophy in Autobiography*, edited by William E.Mann, 181–218. Oxford: Oxford University Press, 2014.
Troncarelli, Fabio. "«Haecine est bibliotheca…» Manuscript Fragments of Boethius' Library". In *«Sit liber gratus, quem servulus est operatus». Studi in onore di Alessandro Pratesi per il suo 90° compleanno*, edited by Paolo Cherubini and Giovanna Nicolaj, 21–26. Città del Vaticano: Scuola vaticana di Paleografia, Diplomatica e Archivistica, 2012.
Valente, Luisa. *Concreti e dividui. Il lessico filosofico di Gerberto di Poitiers.* Rome: ILIESI, 2022.
Valente, Luisa. "*Ens, unum, bonum.* Elementi per una storia dei trascendentali in Boezio e nella tradizione boeziana del XII secolo". In *«Ad ingenii acuitionem». Studies in Honor of Alfonso Maierù*, edited by Stefano Caroti, Ruedi Imbach, Zeno Kaluza, Giorgio Stabile, and Loris Sturlese, 483–545. Louvain-La-Neuve: Féderation Internationale des Instituts d'Études Médiévales, 2006.
Valente, Luisa. "*Illa quae transcendunt generalissima*: elementi per una storia latina dei termini trascendentali (XII secolo)". In *Metaphysica, sapientia, scientia divina: soggetto e statuto della filosofia prima nel Medioevo (Atti del XIV Convegno della Società Italiana per lo Studio del Pensiero Medievale, Bari 9–12 giugno 2004)*, edited by Pasquale Porro, 217–242 [= *Quaestio*, 5 (2005)]. Leiden: Brill, 2005.
Valente, Luisa. "Names That Can Be Said of Everything: Porphyrian Tradition and 'Transcendental' Terms in Twelfth Century Logic". In *Vivarium*, 45 (2007): 298–310.
Van de Vyver, André. "L'évolution du comput alexandrin et romain du IIIe au Ve siècle". *Revue d'histoire ecclésiastique*, 52 (1952): 5–25.

Van de Vyver, André. "L'évolution scientifique du Haut Moyen-Âge". *Archeion* 19 (1937): 12–20.
Van de Vyver, André. "Les étapes du développement philosophique du Haut Moyen-Âge". *Revue belge de philologie et d'histoire* 8/2 (1929): 425–452.
Van de Vyver, André. "Les oeuvres inédites d'Abbon de Fleury". *Revue bénédictine* 45 (1935): 125–169.
Van Winden, J. C. M. *Calcidius on Matter. His Doctrine and Sources*. Leiden: Brill, 1959.
Ventura, Iolanda. "*Aristoteles fuit causa efficiens huius libri*: On the Reception of Pseudo-Aristotle's *Problemata* in Late Medieval Encyclopaedic Culture". In *Aristotle's Problemata in different Times and Tongues*, edited by Pieter De Leemans and Michèle Goyens, 113–144. Leuven: Leuven University Press, 2006.
Verbist, Peter. "Abbo of Fleury and the Computational Accuracy of the Christian Era". In *Time and Eternity: The Medieval Discourse*, edited by Gerhard Jaritz and Gerson Moreno-Riano, 63–80. Turnhout: Brepols, 2003.
Verbist, Peter. "Abbon de Fleury et l'ère chrétienne vers l'an mil". In *Abbon de Fleury. Philosophie, science et comput autour de l'an mil. Actes des journées organisées par le Centre d'histoire des sciences et des philosophies arabes et médiévales, CNRS/EPHE/Université Paris* 7, edited by Barbara Obrist, 61–93. Paris and Villejuif: CNRS, 2006.
Verbist, Peter. *Duelling with the Past. Medieval Authors and the Problem of the Christian Era (c. 900–1135)*. Turnhout: Brepols, 2010.
Vezin, Jean. "Leofnoth. Un scribe anglais à Saint-Benoît-sur-Loire". In *Codices Manuscripti*, 3 (1977): 109–120.
Wallis, Faith. "Introduction". In Bede. *The Reckoning of Time*. Liverpool: Liverpool University Press, 2004.
Wallis, Faith. *Oxford St. John's College 17: A Medieval MS in Context*. PhD. diss., University of Toronto, 1985.
Wallis, Faith. "What a Diagram Shows: A Case Study of Computus". In *Studies in Iconography*, 36 (2015): 1–40.
Warntjies, Immo. "A Newly Discovered Prologue of AD 699 to the Easter Table of Victorius of Aquitaine in an Unknown Sirmond Manuscript". *Peritia*, 21 (2010): 255–284.
Warntjies, Immo. *The Munich Computus: text and Translation. Irish computistics between Isidore of Seville and the Venerable Bede and its Reception in Carolingian Times*. Stuttgart: Steiner, 2010.
White (Peden), Alison. *Glosses composed before the Twelfth Century in Manuscripts of Macrobius' Commentary on the* Somnium Scipionis, 2 vols. Unpublished PhD diss., Oxford: Oxford University, 1981.
Willer, Egil, A. "Henologie". In *Historisches Wörterbuch der Philosophie*, edited by Ritter, Joachim, 13 vols. Darmstad: Wissenschaftliche Buchgesellschaft, 1974, col. 1059, III vol.
Williams, Burma P. and Williams, Richard S. "Finger Numbers in the Greco-Roman World and the Early Middle Ages". In *Isis*, 86/4 (1995): 587–608.
Yeldham, Florence. "Notation of Fractions in the Earlier Middle Ages". *Archeion*, 8/3 (1927): 313–329.
Zambon, Marco. *La ricerca della felicità (Consolazione della Filosofia III)*. Venice: Marsilio, 2011.
Zhmud, Leonid. "Greek Arithmology: Pythagoras or Plato?". In *Pythagorean Knowledge from the Ancient to the Modern World: Askesis, Religion, Science*, edited by Almut-Barbara Renger and Alessandro Stavru, 321–346. Wiesbaden: Harrassowitz, 2016.
Zironi, Alessandro. *Il monastero longobardo di Bobbio. Crocevia di uomini, manoscritti e culture*. Spoleto: Fondazione Centro italiano di studi sull'alto Medioevo, 2004.

Index of Manuscripts

Manuscripts referenced in the endnotes have not been indexed.

Index of Names

Abbo of Fleury and the endnotes have not been indexed.

For Product Safety Concerns and Information please contact our EU representative GPSR@taylorandfrancis.com Taylor & Francis Verlag GmbH, Kaufingerstraße 24, 80331 München, Germany

Batch number: 10397794

Printed by Printforce, the Netherlands